coconut kitchen

Text © 2015 by Meredith Baird
Photographs © 2015 by Greer Inez

Published by Familius LLC, www.familius.com

Familius books are available at special discounts for bulk purchases for sales promotions, family use,
or corporate use. Special editions, including personalized covers, excerpts of existing books, or books
with corporate logos, can be created in large quantities for special needs. For more information, contact
Premium Sales at 559-876-2170 or email specialmarkets@familius.com.

Library of Congress Catalog-in-Publication Data

2015931057
ISBN 9781939629722

Edited by Sarah Echard
Cover and book design by David Miles

Photo credits: Cover photograph by Ninell from Shutterstock.com; infographic art adapted from botanical
print from Kohler's Medizinal Pflanzen circa 1883; pages 6, 17, 20, 62 from Shutterstock.com
All other photography by Greer Inez and Meredith Baird

10 9 8 7 6 5 4 3 2

First Edition

Printed in China

MEREDITH BAIRD

coconut
kitchen

Cooking with Nature's Most Beautifying Superfood

FAMILIUS

contents

introduction

When I was given the opportunity to write a book about coconuts, my reaction was an enthusiastic *Yes!* I love coconuts, and, fortunately, they are worth getting excited about. I did not grow up consuming much coconut in any form. Coconuts were a tropical, festive, and foreign ingredient in my South Carolina home.

I remember buying a mature coconut with my German grandmother at the supermarket—Granny Grover was always up for an adventure. We took the coconut home and literally hacked at it with a machete. Fortunately, no limbs were lost, but most of the water was; by the time the coconut was open, the meat was shredded and muddled with dirt. We were over it. We didn't come back with much to eat, but perhaps some repressed anger was expelled. I didn't really consider the coconut for many years after that.

It was during my time studying raw foods that I started to hear more about coconuts. Recipes called for young coconut—what on Earth was that? Coconut water? Coconut oil? This was before coconut products were ubiquitous on the health food market shelves. Thai Kitchen® coconut milk was about all I could find. You could not buy bottled coconut water that was any degree of fresh, and young coconuts had to be purchased online. Needless to say, the whole coconut element of my newly found culinary passion was a bit intimidating.

It wasn't until I attended culinary school and specialized in raw foods that I really started to understand how to use the coconut. Young coconuts were like gold. Almost all of the most exciting recipes included some form of coconut. Times have changed, and coconut products are now more readily accessible. Of course, purchasing organic and fair-trade versions of these products is important to guarantee their quality. Fortunately, many governments are imposing regulations to ensure that coconut growers cultivate this plant responsibly.

My goal in writing this book is not to laud anything as perfect. I only hope to bring a greater and more simplified understanding to the power of the coconut. Literally every part of the coconut has a use, whether in a culinary application, as a resource, or as a material for a variety of products. My personal experience with coconuts is strictly in beauty, wellness, and culinary art. I've learned so much about my favorite little superfood while writing this book. I am inspired by nature's creation of such a useful, beautifying, and flavorful plant. At the very least, I hope you will be inspired to incorporate some element of the coconut into your life.

My sister once asked me what food I would take to a desert island, and I responded, "The coconut." She said, "Well, they are already there." Problem solved.

coconut kitchen

The coconut is an amazing food and ingredient. I view the coconut as a bit of natural magic—its many health-promoting properties and uses serve as an example of how nature provides. When I look at the coconut, I can't help but think how blessed we are and how the need for processed foods should really be obsolete.

Our culture's approach to food is often very backwards. Natural ingredients are processed or adulterated with preservatives until the end products are far from how nature intended for them to be consumed; they are turned into something foreign to the environment and to our bodies. Such has certainly been the case with the coconut. Coconut oils were "hydrogenated" to increase oxidative stability and to give coconut oil, and products made with it, a longer shelf life; however, hydrogenated coconut oil is a substance far from the pure product. Hydrogenated coconut and palm oils have often been considered the main culprits of heart disease and other degenerative illnesses. Fortunately, with properly executed and well-documented research, we are starting to understand that pure coconut oil is not the villain it was once made out to be. The coconut and all of the ingredients derived from it should be embraced. There are so many uses in both the culinary and skincare realms.

I am not a nutritionist, a scientist, or an expert in the coconut domain (although perhaps more so after writing this book). My job is to read, write, and research to the best of my ability. All of the information in this book comes from reputable sources and the most acknowledged experts in this field. *Coconut Kitchen* is a reduction of my studies, and I've tried to present you with the most accurate information possible.

My only claims to expertise are my abilities to be creative and to craft delicious recipe concoctions. The coconut just happens to be the star of this show. I've enjoyed using coconuts and their versatile character to produce delicious food. They truly are one of nature's most unique foods. Nothing that I say is intended to suggest that adding coconut into your diet can solve all of your health problems.

But I believe that eating any food in its most natural form is beneficial to our health and wellness. Cooking or preparing food in general is a meditative and fulfilling process. In our daily lives, we are often so far removed from our food that it is nice to take a step back and reflect on what it means to eat healthfully and to feel healthy.

The majority of these recipes are raw, vegan, and gluten-free. I make no assumption that this is how you should eat 100 percent of the time. One of the many beauties of the coconut is that it is very conducive to creating recipes within the restrictions of these dietary preferences—and raw vegan food happens to be my culinary specialty.

I am writing this book to serve as an inspiration to live a healthier, happier, and more vibrant life. In my opinion, a cookbook should not be viewed as a recipe for perfection; it should serve as a recipe for organic inspiration. I encourage you to look at each dish as a thought, or a suggestion, of a possible course of action. We all live in different locations and have different ingredients available to us. Please take these recipes and customize them to your own liking—and have fun!

getting
to know
the coconut

"He who plants a coconut tree plants food and drink, vessels and clothing; a home for himself and a heritage for his children."

—SOUTH SEAS SAYING

The coconut is a fascinating plant. It is beautiful and unique in many ways. The image of a coconut palm instantly conjures up visions of tropical vacations, sunshine, and relaxation for many of us. There are very few trees, if any, that have quite the same iconography; the coconut palm is rivaled only by the Christmas tree in its visual familiarity.

The coconut is one of the few foods that we associate with the plant that bears it. The coconut palm is unique among plants because every bit can be used. From the roots to the fruit, it is a valuable resource and has provided economic support and nourishment to many cultures throughout history.

Oddly enough, the coconut palm is not technically a tree because it has no bark, branches, or secondary growth. Botanically speaking, the coconut palm is a perennial, and the trunk is the stem. However, it still warrants the name "the tree of life." Coconuts provide food, drink, fuel, fibers, utensils, skincare products, furniture, and more to cultures around the world.

Dissecting the coconut, both figuratively and literally, gives us a glimpse into one of nature's most useful and inspiring foods.

IS THE COCONUT A FRUIT, A NUT, OR A SEED?

The coconut is generally labeled a tropical fruit. This is debated as a misnomer because the coconut "fruit" is also considered a nut. Technically, the coconut is a specific type of fruit called a *drupe*. There are many nuts that are fruits, and the coconut falls into that category. A drupe is a fruit whose seed is enclosed by a hard, stony covering. Other drupes include peaches, olives, mangoes, plums, and cherries. Coconuts, as well as all other drupes, have three layers: the outer skin layer (exocarp), the fibrous husk layer (mesocarp), and the hard, woody layer that surrounds the seed (endocarp). The coconut meat and flesh make up the solid endosperm of the coconut, while the coconut water is the liquid endosperm of the coconut. The parts of the coconut that we consume are essentially the interior of the seed. The coconut that is available in stores does not resemble the coconut that grows on the coconut palm tree because the exocarp and the majority of the mesocarp have been removed and what is visible is the endocarp. As the coconut matures, the endocarp becomes brown with a thick husk. This is the image that is typically associated with the coconut.

As the coconut matures, it loses water, and the interior flesh becomes thicker and more solid. Mature coconut flesh is delicious and versatile, but its water is much less desirable and sweet. Meat harvested at this stage is often dried to form what is known as the *copra*, which is then squeezed or pressed to extract coconut oil. Coconut oil, coconut flakes, and coconut butter are all extracted from mature coconut meat.

A young coconut is green on the exterior, and the flesh is soft and jellylike inside. Sometimes young coconuts are referred to as *tender-nuts* or *jelly-nuts*. The water content of a young coconut is much higher and more flavorful than mature coconut water. The young coconuts that are typically found in stores have a white, cone-like shape; this is because the green husk has been shaved off to make transportation easier. Unfortunately, when a young coconut is shaved, the white husk starts to brown. To prevent browning, some companies dip the coconuts in harmful chemicals and fungicides. Some young coconuts are in transit for up to two months before they reach the United States and are a far cry from fresh. It is very important to seek organic and high-quality young coconuts.

There are companies that manufacture young coconut meat with integrity and extract it freshly at the source (without chemical or heat treatment). The coconut meat and water are then frozen and shipped directly to you. Although frozen isn't the same as fresh, it is more economically sound than shipping heavily sprayed and bulky coconuts around the globe.

WHERE DOES THE COCONUT COME FROM?

Coconut palms occur naturally in the tropics, particularly in the southern Pacific Ocean. They like warm, humid climates and grow best in temperatures between 80 and 86 degrees, with full sun, and in sandy and aerated soils with a fresh and ample supply of groundwater. Coconuts are not frost resistant and when exposed to cold can develop fatal fungal infections. They are rarely found in climates that dip below 70 degrees. Despite their sensitivity to the cold, coconuts are very resistant to the harsh conditions, strong winds, and blazing sun found in tropical environments.

The top five regions in the world for coconut production are Indonesia, the Philippines, India, Brazil, and Sri Lanka, although all tropical regions have some form of coconut production. Growing regions are found as far north as Hawaii and as far south as Madagascar.

Unlike other fruit seeds, coconut seeds are not transferred by fauna; the seed is too large for any animal to consume or disperse in its whole form. Historically, the coconut plant was spread by water and seafaring people until it was later commercially cultivated. The hollow, water-filled fruit is water resistant and buoyant, able to float from island to island and coast to coast without rot or damage. It is said that coconuts can travel for up to 110 days and still be able to germinate, giving them ample time to disperse.

Scientists have identified more than sixty species of the cocos palm that have existed throughout history, but today there is only one species: the *Cocos nucifera*. Spanish settlers called the fruit *coco*, meaning "monkey face," because of the three indentations that form a face on the bottom of the coconut.

THE COCONUT PALM

Coconut palm varieties are defined by two characteristics: dwarf and tall. The tall variety is planted for both household and commercial uses and grows to a height of twenty to thirty meters. Coconut palms are slow to mature. It can take six to ten years for a coconut to bear fruit and as long as fifteen to twenty years for it to reach peak production. However, they are long lived and can be used for production for sixty to seventy years. Dwarf palms are believed to be a genetic mutation of the tall palm and grow to a height of eight to ten meters. They begin producing fruit around year three but will bear fruit for only up to around twenty years. Once the coconut fruit emerges, it can take about a year for the fruit to fully ripen and mature.

Coconuts grow in bunches and ripen at about the same time. One fertile palm can yield as many as seventy to one hundred fruits per year or as low as thirty to forty fruits per year. When the fruit first emerges, it contains mostly water (young coconut), and over time, the water converts to the white meat that lines the shell. You can tell the maturity of a coconut before it is harvested by the color of the husk (exocarp). Young coconuts are a bright green color. As the coconut begins to mature, the husk starts to turn brown. The coconut is fully mature when the exocarp is solid brown throughout. It is much easier to tell a fully ripe coconut from a young coconut that is good for harvest because the young coconut starts out green and remains green for six to seven months.

If harvesting young coconuts, the coconut's development while on the tree needs to be monitored from flowering to fruit. Another way to tell whether a coconut is mature or young is by the water content. If the coconut is fully mature, there will be very little to no sound of water inside. A young coconut that is ready for consumption will feel and sound like it has ample water content but will be slightly muffled by the development of the flesh on the inner walls.

HARVESTING

Harvesting coconuts is typically done by climbing the stem of the palm with a rope ring around the feet or ankles of the climber or by using a ladder. In some regions, southern pig-tailed macaque monkeys are actually trained to harvest. Once at the top, the climber tests the maturity of the coconut and, using a knife, cuts the bunch at its base, letting the coconuts fall to the ground. If the nuts are tender, or if the ground beneath the tree is particularly hard, the coconuts are collected and carefully lowered instead.

When harvesting, it is important for the climber to not cut any green leaves, because this will affect the yield of the palm. A harvest can happen in varying periods throughout the year and depends on the location of the palms. If the coconut palm is healthy and fully mature, production harvests can occur on a monthly basis.

There is an indisputable fad factor that comes along with many of the coconut products on the market, and there is nothing fresher than a coconut directly from the tree, but as popularity grows, so does the awareness of the unique versatility found in the coconut. There are many companies that are harvesting, producing, and manufacturing coconuts in a sustainable way. As with any agricultural product, it is important to know where your food comes from and to not take advantage of its availability.

FOOD AS MEDICINE

This book will discuss many culinary and practical uses of coconut, but perhaps the best part about consuming coconut is its incredible medicinal and health properties. Coconut oil is one of the first oils historically to be used as both food and medicine. Traditional medicine and Ayurvedic literature have recognized both the nutritional and culinary value of the coconut for centuries.

The coconut is classified as a "functional food" because its benefits extend well beyond its nutritional data; it is a superfood indeed. In the following chapter, the specific benefits of adding coconut into your diet in its various forms are broken down; this overview of the coconut's documented health benefits will show that it is an incredibly powerful food and medicine. Due to the coconut's potent antiviral, antifungal, and antimicrobial properties, it can be used to treat an endless amount of illnesses and ailments with extreme efficacy.

The studies that back up the benefits of adding coconut to your diet are extensive. The most comprehensive resource on the benefits of coconut is *The Coconut Oil Miracle* by Bruce Fife.

Eating a healthy diet means consuming foods that improve your overall wellness and sense of well-being. Despite its medicinal uses and culinary versatility, the coconut has suffered from a bad reputation and has gone in and out of acceptance. Fortunately, due to ongoing research and documentation, some of what was once considered a negative is now understood as a positive. As with most things, all products are not created equally. Hydrogenated or fractionated coconut oils are not the same as pure, pressed virgin coconut oil. Canned coconut milk with stabilizers is not the same as homemade coconut milk. Coconut water from a pasteurized can with sugar added is not the same as fresh coconut water. When choosing foods for their flavor and health benefit, you must consider the source and the production process.

BENEFITS OF COCONUT

Coconut can:

- increase energy
- lower cholesterol
- aid in weight loss and boost metabolism
- improve lung function and treat asthma
- provide heart and arterial protection
- help heal skin abscesses, infections, and rashes
- help with infections of all types
- improve immunity and overall wellness
- act as an antibiotic
- improve eczema and psoriasis
- improve dry scalp
- help heal sore throats
- help alleviate menstrual cramps
- improve thyroid conditions
- combat chronic fatigue and IBS

products

"Let food be thy medicine
and medicine be thy food."

—HIPPOCRATES

This chapter breaks down the various products that are commonly extracted from coconuts. The coconut has so many uses and applications that it is impossible to include them all. Given the title *Coconut Kitchen*, my efforts are focused mainly on the health benefits of the ingredients that are used in culinary art.

COCONUT WATER

"Dew from the heavens." —Hawaiian saying

Coconut water, or coconut juice, is sometimes confused with coconut milk, but these are two entirely different liquids. Coconut water is the liquid from a young, immature coconut. Coconut milk in its purest sense is coconut meat blended with water or coconut water. One has the consistency of water, while the other has the consistency of a thick cream or milk. They are both incredibly delicious and versatile.

Until fairly recently, coconut water was a bit of a mystery outside of the tropics, and it wasn't available in grocery stores. Although coconut water is now widely available, many of the bottled versions on the market are still highly processed, pasteurized, and sweetened, and they are incomparable to fresh coconut water. There is nothing on earth like coconut water directly from the fruit, but because many of us don't have coconut trees available to us, there are companies that source and bottle fresh young coconut water. See the resources section for brands that bottle coconut water with integrity.

E. Coconut Leaves

- thatching for roofs and fences
- baskets, mats, and woven goods
- kindling
- brooms

F. Coconut Trunks

- lumber to make houses
- furniture
- posts and power and telecommunication poles
- floor tiles
- activated carbon

G. Coconut Roots

- medicinal mouthwash
- medicine
- dyes
- toothbrushes

A. Coir of Coconut

The coir of the coconut is the natural fiber extracted from the husk of the coconut. The coir is the fibrous material found between the internal shell and outer layer of the coconut.

- gardening
- floor mats
- brushes
- ropes and string
- mattress stuffing

B. Coconut Shells / Husks

- skincare products for exfoliation
- bowls/ utensils
- simple musical instruments
- charcoal source

C. Coconut Meat

- coconut oil
- coconut milk
- coconut flakes
- copra

D. Coconut Water

- coconut water to drink
- coconut wine
- coconut vinegar

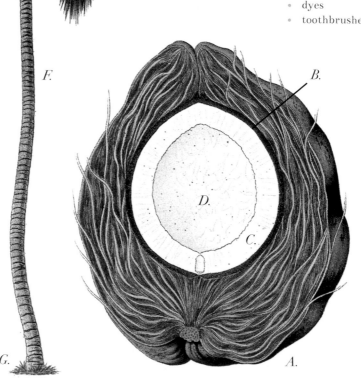

Cocos nucifera

Coconut water boasts a wide range of benefits that can be attributed to its rich mineral density. The high mineral content of coconut water derives from the terroir of the seaside with its salty, sandy, mineral-rich soil. Minerals like calcium, magnesium, and potassium are absorbed by the root system of the coconut palm and transferred into the sweet juice. Coconut water contains more potassium than a banana, making it an excellent sports drink that is far superior to other products on the market. All of the trace minerals of coconut water are in the form of electrolytes, making it unrivaled by tap water for immediate hydration. After a long run or an intense sweat, there is nothing that compares to the satisfaction that comes from a glass of coconut water.

Studies have also discovered that coconut water contains cytokinins, a plant growth substance that promotes cell division. Cytokinins are known for delaying the aging of plant leaves. The analysis of cytokinins, which represent a major group of phytohormones, is an important area of research in plant science. Cytokinins play major roles in cell division, chlorophyll formation, differentiation of plant tissues, seed germination, bud formation, and reproductive development. In addition to the above plant-related roles, some derivatives of cytokinins could potentially be useful for reducing the growth of some types of mammalian tumors.[1] Due to the coconut's cytokinin activity, applying coconut water externally may trigger significant antiaging and anticarcinogenic properties that in turn help balance pH levels, minimize skin aging, and keep skin tissue and cells hydrated and strong.

Not all coconut water is created equally, though. The most desirable coconut water comes from fresh young coconuts. But even young coconut water can be sour. Coconut water is only fresh if it tastes sweet and smells good. Pink coconut water is not desirable in freshly opened young coconuts, as it means that the coconut has turned bad. However, if you are purchasing high-quality, high pressure–pasteurized bottled coconut water, pink water is considered OK. Some bottled coconut waters turn pink over time due to naturally existing antioxidants and phenols interacting with light. This discrepancy is part of a natural change due to processing. Coconut water with a purple or grayish tint should be avoided.

Coconut water is also desirable because, compared to other fruit juices, it is very low in sugar. Sweetness is most often associated with a high sugar and calorie content, but in the case of a young coconut, this is not the case. A cup of young coconut water contains about fifty calories and six grams of sugar. Most other fruit juices contain twice that.

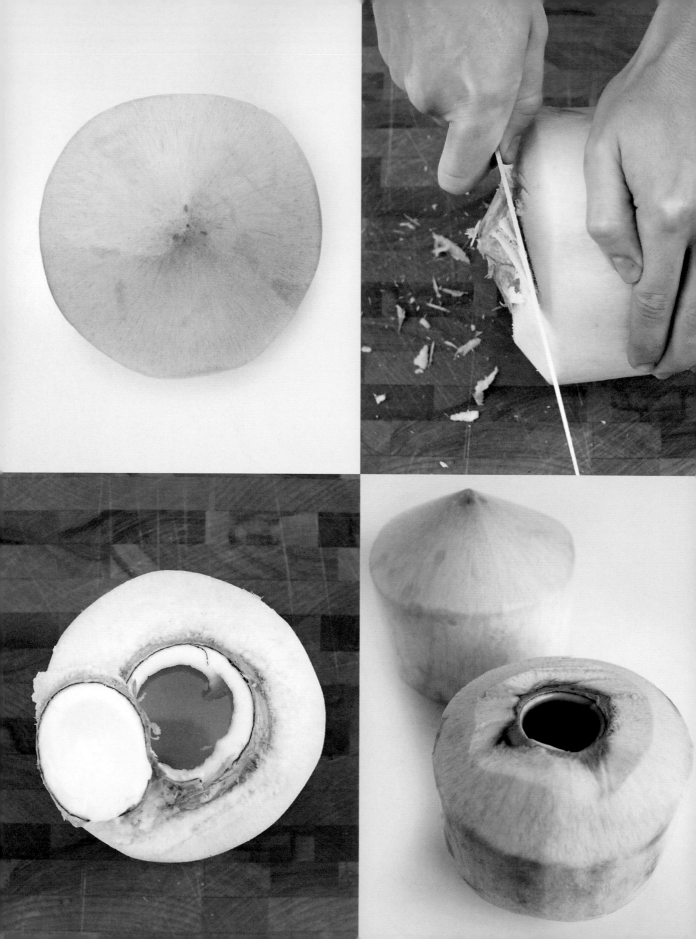

HOW TO OPEN A COCONUT

The only trick to consuming fresh coconut water is breaking the hard shell to access the interior. Young coconuts are much easier to crack than fully mature coconuts. For the purpose of this book, we will focus on opening young coconuts. I prefer a method that is slightly cleaner and less aggressive than hacking at the coconut with a machete or cleaver.

In order to open a young coconut, you will need a hard, steady surface and a large, sharp chef's knife.

1 Place the coconut upright (with the tip on top) on a hard, flat surface that will not be damaged by repeated pounding.

2 Hold the coconut steady with your non-dominant hand so that you have a good grip.

3 Using the chef's knife, shave away as much of the husk on top as possible.

4 Tap the shaved surface with heel of the chef's knife to determine if there are any weak or thin spots. Focus your hacking on this spot.

5 Making sure to keep your non-dominant hand out of the way, make strong whacks with the blade of the knife around the top of coconut in a circular motion. The weak spot will crack.

6 Once you have made a small crack in the coconut, use the heel of the knife to pry open the top and drain the water.

7 You should have an open "lid" that is about two inches in diameter. If the opening of the coconut is too small, now is the time to aggressively whack it in half.

8 Once the coconut is open, scrape out the meat using the back of a spoon.

Practice makes perfect, and some coconuts are much easier to open than others.

COCONUT MEAT

Coconut meat is the interior flesh of the coconut. Most people are somewhat familiar with this ingredient, although for many people, the difficulty of "cracking the coconut" is an inhibiting factor to consuming it regularly. There are two different types of meat: young coconut meat and mature coconut meat. In Ayurvedic tradition, the coconut meat is divided into three types or stages— *baal*: tender or young coconut, *madhyam*: half-mature coconut, and *pakva*: fully mature coconut. Each stage has markedly different characteristics and uses. The young coconut has the highest water content and soft, gelatinous flesh. The half-mature coconut is what we value most for the flesh. It is impossible to distinguish a very young coconut from a half-mature coconut—these are both sold in the market as young coconuts.

In the wild, a young coconut is large and light green. When you see whole young coconuts in the market, they have typically been shaved down and are white with a little pointed top hat. Fully mature coconut meat comes from the furry brown coconuts commonly pictured in the mind's eye, the "monkey-faced" coconut. Fully mature coconuts contain the least amount of water, and their flesh is the most dense and chewy.

Young coconut meat is the prized flesh that will be used the most throughout this book. As a culinary tool, it is invaluably versatile. Young coconut meat is lighter in flavor, density, and calories than mature coconut meat. The texture is gelatinous. Depending on the age of the coconut—and again, this is virtually impossible to tell from the exterior—the meat may be slightly firm and flesh-like or, in a very young coconut, virtually nonexistent. To determine whether the meat is good, look at the color. Bright white flesh is the best, whereas a grayish or purple tint is undesirable.

It is difficult to determine quantities of coconuts for a recipe because the yield of meat within each coconut varies greatly. My best advice: if you are not using prepared young coconut meat, buy one or two more coconuts than you plan on using. See the resources section of the book for a recommendation of a prepared brand of young coconut meat that is organic and is harvested fresh and frozen directly from fair-trade coconut farms in Thailand. If you don't have the time or wherewithal to get your hands too dirty, purchasing frozen young coconut meat is an excellent option.

Not only the appearance but also the nutrition of a coconut changes as it matures. The texture of mature coconut meat is crisp and satisfying, similar to a carrot. This density in fiber is combined with a higher fat and calorie content.

Mature coconut is perhaps a little less versatile—or, at least, it is used in more expected applications. That isn't to say that it isn't incredibly delicious.

Shredded mature meat makes a fantastic addition to salads, soups, rice dishes, vegetable dishes, coconut milk, desserts, and pastries. I personally love to snack on mature coconut meat and find it excellent for efficient digestion.

Young coconut meat is much less dense and, cup for cup, the calorie and fat content is less than the content of mature coconut meat. As you will find throughout the book, coconut provides good fats and calories. Both young and mature coconuts are delicious parts of a balanced diet.

COCONUT FLAKES

Dried coconut flakes that are commonly found in grocery and health food stores are made from dehydrated mature coconut meat. When buying coconut flakes, make sure to buy organic and unsweetened flakes. Most brands of coconut flakes found at the supermarket are highly processed, sweetened, and full of preservatives.

Coconut flakes are similar to mature coconut meat in their nutrition content, but the water content is reduced from approximately 50 percent to 2 percent. Coconut flakes are easier to handle than fresh coconut because they come prepared and have a much longer shelf life. The flakes are an easy, versatile, and more readily accessible way to use the coconut. The flakes are high in fiber and contain high-quality vitamin and mineral content. They will be used in various recipes throughout this book.

Perhaps the most exciting way to use coconut flakes is in learning how to make homemade coconut milk. Making your own coconut milk is easy and virtually free of any preparation or mess. Canned coconut milk is questionable; try to avoid canned foods whenever possible. Making fresh coconut milk opens up a whole new world. Dried coconut flakes or shredded, fresh, mature coconut meat can be used to make coconut cream or milk.

COCONUT OIL

Extracted coconut oil is the most concentrated form of coconut and is also the most potent and medicinal. Coconut oil is the powerhouse of the coconut. Coconut oil has long been recognized in Ayurvedic tradition for health and healing, and most of the South Pacific population has used coconut oil for both culinary and medicinal purposes for thousands of years. Many of these populations enjoy great health and longevity and seem to avoid the degeneration that is seen in cultures that eat more modern diets.

Coconut oil has not been traditionally consumed in the Western diet and was undeservingly given a bad rap. This was mostly due to highly processed and hydrogenated forms of coconut and palm oil that were used in the manufacture of already nutritionally void foods. As detailed in an upcoming section, the type of saturated fat found in coconut oil is markedly different from the type of saturated fat found in animal sources. Consuming coconut oil is one of the easiest and most effective ways to get the benefits of coconut into your diet.

EXTRACTION

There are four different methods of extracting coconut oil:

- fermentation
- expeller pressing
- cold pressing
- centrifuged extraction

Fermentation is the least common and least consistent method of extracting oil from fresh coconut meat. Coconut cream is pressed from fresh coconut meat, and the resulting cream is then fermented in large vats. The fermentation process breaks down the cream into layers of protein solids, liquids, and oils. The oil is separated and filtered from the other layers. Because the oil's moisture content is high, it must be boiled before packaging to remove the excess moisture. Boiling the oil destabilizes it and reduces the shelf life. This process produces the lowest-quality oil.

Expeller-pressed coconut oil is the most common form of production. During the expeller-pressing process, the coconut meat is made into copra, a form of dried meat, then rapidly extracted using heat and pressure. This method produces an oil that is crude and must be refined, bleached, and filtered before using. This product is sometimes referred to as RBD coconut oil (refined, bleached, and deodorized). This is usually the least expensive type of oil—and the least nutritious.

The methodology in the production of cold-pressed coconut oil is very simple. As with expeller-pressed oil, the process begins with copra. The quality of the meat and the way that the meat is dried are important variables in the flavor and integrity of the final result. If the meat is of poor quality, or is burned during drying, the resulting oil can taste rancid. Coconut meat is dried at varying temperatures—some as low as 103 degrees, which will produce a lightly flavored and higher-quality oil, and some at temperatures as high as 170 to 180 degrees, which produces a less flavorful oil. The dried meat is then pressed to extract the oil. When the meat is pressed, some proteins are also extracted, and the resulting oil must be either filtered or decanted to separate it from the proteins.

Pressing is also done at varying degrees and temperatures. Oil that is labeled "cold-pressed" is not always extracted at low temperatures. The temperature at which the oil is extracted and the temperature and quality of drying are key factors in the flavor of the finished product. There is wide variety in the quality of cold-pressed oils, so it is important to choose a manufacturer that is transparent about their practices.

Oil manufactured through centrifuged extraction is made from the cream of freshly pressed coconut meat. The cream is about 40 percent oil. Through a low-temperature centrifugal extraction process, the oil is separated from the solids. The result is a very mild, high-quality oil with regularly consistent results. This oil should always be labeled "virgin" or "extra virgin" and is considered raw. This is the highest-quality and most nutritionally beneficial coconut oil.

WHAT ABOUT COCONUT OIL'S BAD REPUTATION?

Saturated fat was put on the list of foods to avoid because of its link to an increased risk of heart disease and high blood cholesterol levels. Coconut oil was unfairly implicated as it is a purely saturated fat. However, the coconut oil that was used in these studies was far from pure. We now know that not all saturated fats are created equally nor are they all as harmful as they were once considered to be.

Fats are divided into three categories: saturated, monounsaturated, and polyunsaturated. Within these categories there is much variation, and one fat type cannot be labeled as completely healthy or unhealthy. In typical Western diets, we are encouraged to eat a diet higher in polyunsaturated fats; these fats are considered healthy and include most plant-based oils like olive, nut, and seed oils as well as fish and algae. Advice to substitute polyunsaturated fats for saturated fat has been the key message in worldwide dietary guidelines for coronary heart disease risk reduction. However, these studies have been inconclusive, and in some cases, substituting dietary polyunsaturated fat for saturated fat actually increased the rates of heart and cardiovascular disease. These studies highlight the importance of eating a wide variety of high-quality, whole food–based fats including coconut oil.

Coconut oil is one of the most highly saturated fats at 92 percent saturation. This is higher than butter, beef fat, and lard. Highly saturated fats and oils are solid at room temperature, whereas polyunsaturated and monounsaturated fats are liquid. You might wonder: *How can this be healthy?* The advantages of saturated fat are that it is more stable and less susceptible to oxidation and free-radical damage than other fats, making it a purer source of nutrition. Consuming oxidized, highly processed, and poor-quality oils is, I believe, unnatural to our bodies and damaging to our health. The stabilized nature of saturated fat provides a high-quality source of energy and nutrition. Saturated fat acts as an efficient carrier of the fat-soluble vitamins A, D, E, and K and also aids in mineral absorption. Our brains are made up of about 60 percent saturated fat. Healthy fat is imperative to include in a healthy diet.

The length of the fatty acid chain also plays a role in determining the effects of fats on our health. There are three acid chain lengths: long-chain triglycerides (LCTs), medium-chain triglycerides (MCTs), and short-chain triglycerides (SCTs). In both plant-based fats and animal fats, the most common triglyceride is long-chain or LCT. Most of the fats and oils that we consume come in the form of long-chain triglycerides. Coconut oil is unique in that it is composed of medium-chain triglycerides or medium-chain fatty acids (MCFAs).

MCFAs are small and easy to digest; they are therefore more easily absorbed by the body than long-chain fatty acids. MCFAs do not require bile acids for digestion and are sent directly to the liver to be converted to pure energy. Easy digestion generally means easy energy conversion. What makes the coconut unique is that it is one of the only and one of the purest sources of MCFAs—other sources include palm kernel oil and, to a lesser degree, butter. MCT oils are now sold in health food stores, but consuming MCT in a whole form, like coconut oil, is best.

Coconut oil is also rich in lauric acid. Lauric acid is a medium-chain fatty acid found mostly in coconut oil, palm kernel oil, cow's milk, goat's milk, and breast milk. Lauric acid is antifungal, antibacterial, and antiviral. Lauric acid gives coconut oil its incredible immune-boosting and infection-fighting properties. The benefits of lauric acid are just now starting to be recognized, and coconut oil is one of nature's most concentrated sources of this healthy fatty acid. Coconut oil contains caprylic acid and capric acid, which also contribute to coconut oil's powerful antifungal, antiviral, and antibacterial properties.

The impact of coconut oil on blood cholesterol also seems to have been misunderstood. Although the effects are still being studied, there is some evidence that due to metabolic stimulation, coconut oil may indirectly lower LDL (bad) cholesterol and increase HDL (good) cholesterol.[2] Studies on coconut-eating populations further propel this claim. Populations that traditionally consume large quantities of coconut oil have a very low incidence of heart disease and have normal blood cholesterol levels, and when they change their diets and replace coconut oil with refined polyunsaturated vegetable oils, their risk of heart disease increases.[3]

As the coconut's popularity grows, the studies on the benefits of coconut oil are becoming more extensive and thorough. Saturated fat from natural whole food sources no longer needs to be considered a villain, and removing these foods from our diet might very well be what has contributed to many of our most common health crises. Tapping into the unique powers and simplicity of the coconut is truly exciting.

COCONUT BUTTER

Coconut butter is sometimes confused with coconut oil, but the two are very different. Coconut butter is made by grinding dried coconut flakes to form a creamy spread. It is a relatively new product on the market, but it is very easy to make at home. Included in the next chapter is a coconut butter recipe and recipe variations. You can use coconut butter in place of any of your favorite nut butters—as a snack or on sandwiches.

Consuming coconut butter is an easy way to get the beneficial fat and fiber of the coconut without having to open a coconut to extract the fresh meat. However, coconut butter is not a direct substitution in culinary applications for fresh young coconut meat or for coconut oil.

COCONUT SUGAR

Coconut sugar is the sap extracted directly from the flower of the coconut palm—it does not come from the coconut itself. The sap is evaporated at low heat and is generally a minimally processed form of concentrated sweetener. Coconut sugar has become highly regarded as a great alternative to dried table sugars, brown sugar, or date sugar and as a substitute for liquid sweeteners like corn syrup, maple syrup, or agave. Coconut sugar is lower on the glycemic index than other sweeteners (GI 35) and contains a wider variety of minerals, vitamins, and amino acids. Although coconut sugar is one of the best choices of concentrated sweetener, like all sugar, it should still be consumed in moderation.

The flavor of coconut sugar lends itself very well to substitution. It is naturally sweet and complex with slight caramel and toasted nut undertones. The flavor is not as neutral as agave, but it is more neutral than honey or maple syrup. The substitution ratio is 1:1 when using it in either its dry or liquid form.

CLARIFICATION ABOUT COCONUT SUGAR

Coconut sugar may be presented as:

- coconut sugar
- coconut palm sugar
- coconut sap
- coconut crystals
- coconut nectar (coconut sugar in its liquid form)

Coconut sugar can be a confusing subject because it is known by many different names and has different levels of integrity. Included in the resources are some companies that produce high-quality, minimally processed coconut sugar.

Both coconut sugar and coconut nectar have a wide variety of uses and can be substituted for virtually any other sweetener. This chapter has touched upon the health benefits of using coconut as a sweetener, but selecting the right sugar can be a challenge. This section will clarify the difference in flavor, texture, and usage for each of these products.

The term *coconut sugar* refers to evaporated coconut granules or dry sugar. Coconut sugar is an unrefined sweetener made from the sap of the cut flower buds of the coconut palm. Coconut sugar in this form has a similar taste and texture to brown sugar, cane sugar, or rapadura. The crystals are much softer than white sugar and dissolve very easily into liquids. The flavor of coconut sugar is perfectly sweet but not harsh like other more refined sweeteners. Coconut sugar can be substituted in virtually any recipe that

Coconut nectar

calls for a dry sweetener like brown or white sugar. Products labeled *palm sugar* are technically still coconut sugar if they come from the coconut palm, but that isn't always the case; the sugar is sometimes extracted from other types of palms. This product is not coconut sugar. The distinction between coconut sugar and palm sugar depends on the producer.

Coconut nectar is the maple syrup of the coconut tree. When the tree is tapped, it produces a sweet sap that is then evaporated at low temperatures to produce coconut nectar. Coconut nectar is a slightly viscous liquid with a rich amber-gold color and is full of essential minerals and amino acids. The flavor of coconut nectar is very versatile; it tastes sweet without being cloying and is more neutral than honey or maple syrup. Coconut nectar has a slightly richer and more complex flavor than agave. The only inhibiting factor in using coconut nectar as a liquid substitution is that the rich color can turn lighter foods slightly brown.

Pure Thai coconut sugar is the queen of coconut sugar. This product is slightly harder to find in stores than coconut sugar or coconut nectar, but it can be found online. Thai coconut sugar is derived from the sap of the coconut blossoms and is thickened at low temperatures. The sweet sap surrounds a block of crystallized coconut sugar. Pure Thai coconut sugar is the spun honey of coconut sweeteners. It is so creamy and delicious that it is almost best served on its own. The flavor is slightly richer and more full bodied than the more commonly found coconut nectars. This product can be eaten directly from the container. It also tastes amazing on toast or as a substitute for caramel on ice cream. It can be substituted in almost any application in which coconut nectar would normally be used.

COCONUT AMINOS

Coconut aminos are a soy-free sauce alternative made when coconut nectar is combined with highly mineralized sea salt. The aminos are created by being naturally processed through fermentation and aging. Coconut aminos are richer in minerals and amino acids than soy sauces. Some studies claim that coconut aminos contain up to fourteen times the amount of amino acids as regular soy sauce.[4] The beauty of coconut aminos is that they are gluten-free and often made in small batches to ensure quality and an enzymatically alive product. The flavor is slightly sweet, salty, and less harsh than soy sauce. Coconut aminos can be substituted at a 1:1 ratio in any recipe that calls for soy sauce, tamari, or nama shoyu.

COCONUT VINEGAR

Coconut vinegar is produced from either the sap of the coconut tree or from mature coconut water. Coconut vinegar made from sap is considered to be higher quality and nutrient denser than coconut vinegar made from coconut water and is, therefore, usually more expensive. Sap vinegar is naturally aged in barrels for a period of eight months to a year. The result of this fermentation produces an enzymatically alive, probiotic-rich product that is similar to apple cider vinegar in taste and versatility. Purchasing high-quality fermented and aged vinegars—whether they be coconut vinegars or not—is important. Many commercial vinegars on the market are pasteurized and do not contain the traditional healing properties found in quality vinegars.

COCONUT FLOUR

Coconut flour is made from dry, defatted coconut meat. The flavor is only slightly sweet and very versatile. It can be used in both sweet and savory applications. Coconut flour is gluten free, is high in fiber, and makes an excellent substitute in all gluten-free baking. It does need to be mixed with other flours because the texture is dense and it absorbs moisture easily; too much coconut flour can create a dry finished product. Recipes using coconut flour have not been included in the following chapters because this book's focus is on pure, unheated culinary applications. Coconut flour is great for baking, but it is difficult to use in its uncooked form.

Coconut Meat with Avocado and Gomasio (page 53)

practical consumption

"The discovery of a new dish confers more happiness on humanity than the discovery of a new star."

—JEAN ANTHELME BRILLAT-SAVARIN

To call the concoctions in this chapter "recipes" would be a stretch. This chapter is all about subtle infusions, enhancements, and interesting ways to use the products the coconut offers.

The beauty of the coconut, aside from all of its interesting culinary uses, is that it is perhaps most delicious and most intriguing in its natural form. What tastes better than coconut water fresh from the shell? Fresh, sweet coconut meat, whether young or mature, has a pure flavor that is subtle and intoxicating. The benefits of the coconut can be likened to the benefits of breast milk. It is reasonable to conclude that fit, healthy, glowing islanders may owe some of their vigor to the coconut.

Nature has a way of providing us with what we need based on our location. In harsh temperatures under a blazing tropical sun, the coconut has provided external protection and internal hydration for centuries and has contributed to human adaptation within these environments. There are many foods that we eat that need to be cooked, salted, smoked, or processed in order to make them palatable; the coconut is not one of them. This book makes no presumption that humans can improve upon nature, but this chapter outlines a few ways to mix it up and get creative with the coconut in its most natural form.

coconut water

Coconut water is one of the most hydrating and healing beverages on earth; why change it? Because sometimes, change is refreshing.

coconut water *with* bulgarian rose

Rose is everywhere in the Valley of the Roses in Morocco. Desserts, pastries, and couscous dishes are all gifted with the scent of rose. Rose is so much more than a fragrance or a flower; it is a powerful wellness tool. Rose is especially beneficial for women because it can help relieve menstrual pain and cramping. It also acts as a mild sedative and antidepressant, which is helpful during the emotional sensitivity many women experience during premenstrual syndrome. Consuming rose promotes healthy digestion and detoxification, improves mood, increases energy, and even acts as an internal breath freshener. Rose water is rich in beneficial antioxidants, flavonoids, tannins, and essential vitamins. It is also highly anti-inflammatory, and in combination with coconut water, it makes an elegant medicinal cocktail. The color of this drink is so exotic, and the flavor combination of coconut and rose is natural perfection.

2 cups coconut water
1 ounce Royal Sense® Bulgarian rose water (rose hydrosol) or other
     drinkable rose water
1 teaspoon beet juice (optional, to enhance the color)

Stir all ingredients together. Chill and serve.

Makes 1 serving.

Using Royal Sense® rose water or other high-quality, steam-distilled rose water is recommended to ensure that no additives or chemicals are present. Rose water (or hydrosol) is like rose essential oil but is far less concentrated and can be added by the ounce to any number of foods; it is added regularly throughout this book. Feel free to experiment on your own.

coconut water *with* chlorophyll

Chlorophyll is the green pigment found in plants that, along with sun, serves as the plant's life force. It also gives plants their green color. Chlorophyll closely resembles our own red blood cells and has strong oxygen-carrying capabilities. Liquid chlorophyll drops can be purchased in health food stores and can be used as a light supplement to detoxify, improve digestion, and increase alkalinity. Chlorophyll drops turn coconut water a beautiful shade of green and give it a minty, fresh flavor; chlorophyll CO_2 is also excellent post-workout.

2 cups coconut water
A few drops of chlorophyll

Stir all ingredients together. Chill and serve.

Makes 1 serving.

coconut water *with* jalapeño *and* lime

In this recipe, the sweetness of the coconut perfectly balances out the spicy jalapeño, and the lime brings it all together. It's the perfect afternoon pick-me-up. Jalapeño is great for weight loss because it curbs hunger and raises metabolism. It also contains capsaicin, the chemical compound responsible for the pepper's spicy flavor. Capsaicin is anti-inflammatory, promotes healthy blood flow, and serves as a powerful antioxidant.

2 cups coconut water
1 or 1/2 jalapeño, deseeded and thinly sliced
Juice of 1/2 lime
Pinch of sea salt

Stir ingredients together and infuse for at least 30 minutes or longer, depending on your flavor preference.

Makes 1 serving.

coconut water *with* blueberry, lemon, *and* lavender

This recipe is summer lemonade without the added sweetener. Lavender promotes a sense of well-being and calm and can relieve nervousness and depression; lavender tea is excellent to have before bed. Blueberries and lemons are both high in vitamin C and antioxidants.

1/4 cup lavender tea brew*
2 cups coconut water
Juice of 1 lemon
1/4 cup or a handful of fresh blueberries
1–2 stalks of fresh lavender (for garnish or to enhance the infusion)

To brew lavender tea: steep 1 teaspoon of lavender flowers with 8 ounces of hot water for 5–7 minutes, depending on the strength you prefer.

Stir all ingredients together, lightly crushing the blueberries to release some of their flavor and color.

Makes 1–2 servings.

coconut water *with* citrus *and* ginger

Coconut goes so well with orange. The two flavors together taste like an orange cream dessert. Ginger adds a warming touch that increases immunity and improves digestion. This combination is an excellent way to consume coconut water in the fall and winter months when citrus is in peak season.

1–2 cups coconut water
Juice of 1 orange
1–2 ounces ginger juice, or 4–5 thin slices of fresh ginger

Stir all ingredients together. If infusing the ginger, allow mixture to sit for at least 20 minutes to absorb the flavor.

Makes 1–2 servings.

coconut water *with* turmeric *and* cayenne

This is a potent health tonic. Both turmeric and cayenne are highly anti-inflammatory, are high in antioxidants, boost immunity, and promote well-being. This concoction is designed to provide relief from aches, pains, and headaches. Fresh turmeric is best, but if you can't find it, dried turmeric also works well.

1–2 cups coconut water
1–2 ounces turmeric juice, or 1/2 teaspoon turmeric powder
Pinch of cayenne

Stir all ingredients together.

Makes 1–2 servings.

coconut water *with* beet juice *and* grapefruit

Not only are beets beautiful but they also provide energy and boost stamina. Beets are high in many vitamins and minerals, including potassium, magnesium, fiber, phosphorus, iron, and folic acid. Due to their high amount of boron, beets can boost testosterone production and are considered an aphrodisiac in many cultures. Beets are also good for cleansing and detoxifying. The flavor combination of beets and grapefruit is a natural match in the culinary world.

1–2 cups coconut water
1/4 cup beet juice
1/4 cup grapefruit juice

Stir all ingredients together.

Makes 1–2 servings.

coconut water *with* aloe

It wasn't until writing this book that I really became fond of the aloe plant. Aloe is packed with vitamins, minerals, amino acids, and phytonutrients. It is amazingly beneficial for digestion and can help eliminate constipation; for that reason, I love to have this beverage first thing in the morning to help hydrate and get things moving. Aloe is also anti-inflammatory, antiviral, and antiparasitic. There are so many benefits to adding aloe into your diet. If you aren't growing your own aloe plant, make sure you are buying preservative-free, unsweetened aloe juice.

1–2 tablespoons aloe juice
2 cups coconut water

Stir ingredients together. Serve chilled.

Makes 1 serving.

piña colada *with* coconut water ice cubes

As simple as it sounds, freezing coconut water in ice cube trays is a great little trick. I'm not a big fan of ice in my smoothies, but if you are going to use it, the coconut water ice cube is the only way to go. Not only do you get the extra nutritional benefit but you also get the slight additional sweetness without any added sugar. Coconut water ice cubes are a great addition to many cocktail or "mocktail" beverages.

1–2 cups coconut water
1/4 cup young coconut meat
1/4 cup pineapple chunks
2–4 coconut water ice cubes
Pinch of sea salt

Blend all ingredients until smooth. Garnish with coconut flakes.

Makes 1 serving.

coconut meat

Fresh young coconut meat, mature coconut meat, and dried coconut flakes all have a wide variety of culinary uses, but perhaps they are most exciting on their own. Eating coconut in its purest form tastes amazing. When something tastes so great as is, it's hard to justify blending and altering. Try eating plain mature or young coconut meat in the morning post-exercise for a satisfying and fulfilling boost.

The following recipes are all simple and satisfying ways to consume coconut meat.

A FEW REASONS TO CONSUME COCONUT IN THE MORNING (OR ANY TIME OF DAY)

Coconut:

- improves digestion and absorption of key nutrients, vitamins, and minerals
- supports immune health
- provides a natural source of energy
- enhances physical and athletic performance
- aids in weight loss
- supports thyroid function
- improves skin health and prevents wrinkles

coconut milk *or* coconut cream

Making your own coconut milk is an easy skill to learn. The following is one of the best and most versatile recipes in this book.

2 cups shredded coconut (dried or fresh), or 1 cup young coconut meat
4 cups water, or more or less depending on how thick you would like it
Pinch of sea salt
Scrapings from 1/2 vanilla bean, or 1 teaspoon vanilla extract (optional)
1 tablespoon coconut nectar or coconut honey, or 2 dates (optional)
1 tablespoon lecithin (optional, to prevent separation)

Blend all ingredients until smooth. Run through a fine mesh strainer to remove solids.

Makes 4 cups. Fresh coconut milk lasts up to five days in the refrigerator and can last up to a week if sweetened with honey.

○ *When making this recipe, adjust the ratio of water to coconut according to desired consistency. The less water you use, the thicker and more cream-like the milk will be.*

coconut meat *with* bee pollen

Bee pollen is a holistic remedy known throughout the world for its incredible healing properties. Bee pollen has many of the same health benefits as coconut, and together they make a potent health aid. Bee pollen is an ally against allergies, inflammation, and digestive distress. It also enhances energy and improves cardiovascular health. This combination is the perfect post-workout combo.

1 cup young coconut meat, roughly cut into large pieces
1 tablespoon bee pollen
Pinch of sea salt
1 teaspoon coconut sugar (optional)

Sprinkle bee pollen, sea salt, and optional coconut sugar on coconut meat and serve.

Makes 1 serving.

coconut flakes *with* maca *and* banana

This is a ridiculously simple creation. It's great for breakfast or post-workout. You'll get the idea with this and can basically use any fruit.

Maca is a root native to Peru. It is sometimes referred to as "Peruvian Ginseng." The flavor is slightly earthy and malty, and the root pairs well with fruit or sweeter foods. Maca is rich in vitamins B, C, and E. It is also a good source of calcium, zinc, iron, magnesium, phosphorus, and amino acids. Maca is widely used to promote sexual function in both men and women because it boosts libido and endurance. Maca can also balance hormones and increase fertility. If you are pregnant or nursing, it is best to avoid maca, and if you are serving this recipe to children, substitute cinnamon.

1 banana, peeled
1–2 tablespoons maca
2–3 tablespoons coconut flakes
Pinch of sea salt (optional)

Roll peeled banana in maca and then in coconut flakes. Slice the banana or eat as is. Optionally add a pinch of sea salt.

Makes 1 serving.

coconut meat *with* kimchi

This combination might sound a little odd, but if you like coconut and kimchi separately, you will love the two together. Young coconut meat has a flavor and texture similar to tofu, so it makes a nice substitution. The sweetness of the coconut cuts into the spicy flavor of the kimchi, and it's really quite satisfying. This is the perfect afternoon pick-me-up snack and a good excuse to eat more fermented foods.

1 cup julienned young coconut meat
1/2 cup of your favorite kimchi

Toss ingredients together. Serve with nori, lettuce cups, or your favorite gluten-free bread.

Makes 1 serving.

> *Try mixing this up and scooping it into a lettuce cup or nori. Sauerkraut instead of kimchi is also very good in this recipe.*

coconut meat *with* avocado *and* gomasio

Gomasio is a simple Japanese condiment made of toasted sesame seeds and sea salt. It is very easy to make your own, and you can also find it in most health food stores and markets. I like a version that contains a little bit of seaweed. Gomasio serves as a healthy alternative to traditional table salt. The flavor blends particularly well with coconut and avocado because it isn't as harsh. This combination will keep you satisfied all day long.

1/2 cup sliced young coconut meat
1/2 avocado, cubed
1 teaspoon gomasio
Juice of 1 lemon or 1 lime

Toss all ingredients together. Serve with nori to make a simple wrapped hand roll.

Makes 1–2 servings.

carrot *and* coconut

Mature coconut and carrots have a similar texture and a subtly similar flavor. Both are firm and lightly sweet. For this reason, they are great when served together. You can enhance this dish any way you would like by adding extra spices, herbs, and vegetables. Try adding turmeric and cayenne for an anti-inflammatory and immune-enhancing boost. This recipe makes a great substitution for coleslaw or any other relish-style side dish.

1/2 cup shredded mature coconut meat
1/2 cup shredded carrots
A few teaspoons of sesame or coconut oil
Pinch of sea salt
Pinch of turmeric (optional)
Pinch of cayenne (optional)

Toss all ingredients together. Allow the flavors to marinate for at least 10 minutes before serving.

Makes 1–2 servings.

coconut *and* cucumber

The combination of coconut and cucumber is incredibly hydrating and replenishing to skin. This recipe is great post-workout or on a hot summer day. This dish is also delicious with cilantro, mint, basil, or any herb of your choice.

1 cup julienned young coconut meat
1 cucumber, julienned
A squeeze of lime
Pinch of sea salt

Toss coconut and cucumber together with a squeeze of lime and a pinch of sea salt.

Makes 1 serving.

coconut meat *with* chili *and* lime

In Mexico, this dish is referred to as *cocos preparados*. It is a simple and nutritious combination commonly found in markets and food stands around the country. It is the perfect combination of sweet, salty, sour, and spice. Although young coconuts are traditionally used, I think it tastes great with either mature or young coconut meat.

1 cup young coconut meat or mature coconut meat, sliced in your preferred size
1/2 teaspoon chili powder, or more or less depending on your heat tolerance
Juice of 1 lime
Pinch of sea salt

Toss all ingredients together.

Makes 1–4 servings.

spinach *with* coconut flakes

The flavor of this combination is so simple yet incredibly satisfying. It makes a great base for any salad addition but is excellent on its own. Some of my favorite ways to enhance this dish are with strawberries, fresh avocado, and pumpkin seeds. This recipe leaves plenty of room to get creative.

4 cups spinach
2–4 tablespoons coconut flakes
Juice of 1 lemon
Pinch of sea salt
1–2 teaspoons coconut oil (optional)
Pinch of cayenne (optional)

Toss all ingredients together. Massage the spinach with clean hands to break down the fibers and to marinate it thoroughly.

Makes 2–4 servings as a side dish or 1 serving as a main dish.

coconut oil and coconut butter

Coconut oil and coconut butter are by no means the same ingredients, but they are often used together in recipes or, in simple applications like the ones following, one can be substituted for the other. Coconut oil is the pure oil extracted from the coconut meat, and coconut butter is the whole meat of the coconut pureed into a buttery spread. While both are delicious, coconut oil is perhaps the champion in terms of versatility. Coconut oil can be substituted tablespoon for tablespoon for any oil or butter you would use in standard cooking. Coconut oil has a very high flash point, which makes it an excellent choice for baking, cooking, or sautéing. I could literally write an entire book using coconut oil in recipes. Whether you are baking sweet potatoes, making stir-fries, frying an egg, or popping popcorn, coconut oil is an excellent choice.

coconut butter

This recipe is an invaluable resource to the coconut pantry. Coconut butter is usually an expensive specialty item, but it is so easy to make your own. You'll never have to purchase coconut butter again.

2 cups coconut flakes
1–2 tablespoons coconut oil (if needed)

In a high-speed blender, puree coconut flakes on a high setting to the point that the blender is slightly warm. This process will form the coconut butter. If the flakes are particularly dry, or if there is not enough friction in the blender, add a few tablespoons of coconut oil in order to fully puree the flakes until smooth. The resulting product should have the consistency of peanut butter or almond butter.

Makes approximately 1–1 1/2 cups. Coconut butter lasts in the refrigerator for up to four months.

Top: coconut oil
Bottom: coconut butter

coconut coffee *or* immortality latte

This is the Coconut Kitchen version of Bulletproof® coffee. The Bulletproof® coffee combination offers amazing flavor, an energetic and metabolic boost, and mental clarity. It is traditionally served with ghee or butter, but coconut oil and coconut butter have been substituted in this recipe; the harsh acidity and jittery boost of coffee are eliminated by the coconut oil and coconut butter. It's amazing how blending the oil and butter creates the perfect frothy latte. This beverage tastes like pure nutrient-dense fuel. The oil servings are on the lower side because coffee shouldn't be substituted as a meal—a little bit goes a long way in both flavor and nutrition.

If you choose not to consume coffee, substituting a medicinal mushroom powder is equally delicious and perhaps more nutritious. The combination of reishi powder and cordyceps creates a powerful immune and energy-boosting tonic. Unlike coffee, reishi and cordyceps aid in strength, stamina, and stress relief.

1 cup organic brewed coffee
1/2 tablespoon coconut oil
1/2 tablespoon coconut butter

or

1 cup boiling water
1/2 teaspoon reishi powder*
1/2 teaspoon cordyceps powder*
1/2 tablespoon coconut oil
1/2 tablespoon coconut butter
A few drops of stevia (optional)

*Immortality Organics offers high-quality reishi and cordyceps powder. extracts without any fillers, stabilizers, or synthetic ingredients. Their products are potent and are sourced with integrity.

Blend all ingredients together until smooth and frothy. Sweeten with coconut nectar or stevia if desired.

Makes 1 serving.

coconut butter *with* apple, honey, *and* cinnamon

This recipe will help you increase your morning healthy fat intake and is often on my menu for a simple, satiating morning solution. Spun honey is recommended in this recipe. Cinnamon adds extra immune-boosting and anti-inflammatory properties.

1 apple, chopped
1–2 tablespoons coconut butter
1–2 tablespoons honey
Pinch of cinnamon
Pinch of sea salt

Mix all ingredients in a bowl and eat like cereal.

Makes 1–2 servings.

○   *Try adding a few sprinkles of buckwheat or your favorite granola for a more filling dish.*

coconut butter *and* blueberries

This is the ultimate breakfast when summer blueberries are in season. The recipe sounds so simple, but it is delicious, filling, and packed full of antioxidants.

1 cup blueberries
1–2 tablespoons coconut butter
1–2 tablespoons spun honey

Stir all ingredients together.

Makes 1 serving.

○   *Add bee pollen for an extra energy and protein boost.*

coconut toast

2 slices of your favorite gluten-free or sprouted grain bread
1–2 tablespoons coconut butter or coconut oil
Pinch of turmeric (optional)

This creation is made exactly as it sounds. Spread 1–2 tablespoons of coconut oil or coconut butter onto your favorite gluten-free or sprouted grain toast. Sprinkle with a few pinches of flaky sea salt and it will taste like butter.

Makes 1 serving.

> *If you want to get creative, add a few pinches of turmeric to the coconut butter or oil to give it a light yellow buttery color. Turmeric also gives an added anti-inflammatory boost.*

COCONUT OIL

As mentioned in the introduction to this chapter, coconut oil is a champion of culinary versatility. If you don't have access to fresh coconuts, or if you're not feeling very experimental in the kitchen, I encourage you to at least start trying coconut oil in place of other oils you regularly use in cooking. The health benefits of adding saturated fats back into our diet are only starting to be celebrated; I firmly believe in the quality and stability these fats offer to our diets. The dishes below are some simple ideas to get your culinary juices flowing—all you need to add is a pinch of sea salt.

- Oatmeal with coconut oil
- Baked apples with cinnamon and coconut oil
- Roasted carrots with coconut oil and cumin seed
- Roasted sweet potatoes with coconut oil
- Steamed kale tossed with lemon juice and coconut oil
- Steamed broccoli tossed with coconut oil
- Stir-fry with coconut oil
- Popcorn with coconut oil

ferments

"To ferment your own food is to lodge a small but eloquent protest—on behalf of the senses and the microbes—against the homogenization of flavors and food experiences now rolling like a great, undifferentiated lawn across the globe. It is also a declaration of independence from an economy that would much prefer we remain passive consumers of its standardized commodities, rather than creators of idiosyncratic products expressive of ourselves and of the places where we live, because your pale ale or sourdough bread or kimchi is going to taste nothing like mine or anyone else's."

—*MICHAEL POLLAN*, COOKED: A NATURAL HISTORY OF TRANSFORMATION

The powers of fermentation are no secret. Fermentation is one of the oldest and most treasured forms of food production and preservation. Many of our most common delicacies come from the fermentation and aging process—wine, cheese, sauerkraut, miso, chocolate, coffee beans, and yogurt are all examples of fermented foods that we consume regularly or are familiar with.

When foods were less clean and refrigeration didn't exist, fermentation was a technique that preserved foods and enhanced their nutritional value. Fermentation is the process through which food is exposed to bacteria and yeasts either through deliberate introduction of cultures or by natural exposure to air. There are many different forms of fermentation and culturing. Fermentation is a metabolic process in which beneficial microorganisms eat up the carbohydrates and sugars in food. The result is a preserved product full of beneficial bacteria (probiotics); the active cultures increase digestibility and create unique flavor profiles.

Almost every culture has a fermented food that is unique to their region. Asia has kimchi, miso, soy sauce, tempeh, and kombucha, to name a few. Europeans enjoy sauerkraut and fermented fish. Kefir and various pickled vegetables are common in the Middle East. North America, with its fear of bacteria, overuse of antibiotics, and sterilized diet, consumes the least amount of fermented food of any culture and has compromised the integrity of its cuisine as a result.

Fermented foods increase the beneficial bacteria in our systems by providing a healthy dose of probiotics and digestive enzymes. Fermented foods can improve skin conditions and digestion and increase immunity against bacterial and viral infections. Acne, eczema, yeast overgrowth, and digestive distress can all be a result of an imbalance in your body's microflora; they can also indicate a need to increase the amount of ferments in your diet.

The cultural commonality of fermented foods in our diets for centuries can be taken as a sign that we were designed to eat them. Complexity in flavor is often the result of integrity in product. A one-note wine or cheese does not have the same appeal as versions with more nuances and character. The qualities of terroir, climate, yeasts, bacteria, and molds are all part of an organic symbiosis that makes food flavorful and healthy. When food is homogenized and sterilized, it is broken away from what nature intended it to be, and as a result, much of the nutritional benefit—and flavor—is removed.

Coconut lends itself beautifully to fermentation. Coconut water kefir, coconut cream kefir, and coconut yogurt are all delicious ways to consume fermented foods without consuming dairy or animal products. Once you get the hang of making your own coconut ferments, the process will become very easy and will require little active time. Each of these recipes serves as a base for an endless array of flavor combinations. These recipes are tart and pure, but feel free to make changes if you prefer a little sweetness in the form of fruit or coconut nectar post-fermentation.

Dry kefir granules

coconut cream kefir

It is a common misconception that kefir has to be made with dairy. Milk kefir grain starters can actually be used with coconut milk, rice milk, or other alternative milks. Although milk kefir grain starters themselves are not entirely dairy-free and must be rehydrated in pasteurized dairy milk, kefir made with coconut milk contains only trace amounts of dairy and does not pose a problem for anyone with lactose intolerance.

Coconut kefir is a sour-tasting beverage similar to liquid yogurt. The flavor of kefir is a little more challenging and unfamiliar than yogurt—very tart with a slight hint of yeast. Coconut cream kefir makes a great base for smoothies, milkshakes, ice cream, and even dressings; the flavor is similar to buttermilk. Do not substitute kefir in any heated applications, because heat will kill the cultures.

Although kefir and yogurt do have similar properties and flavor profiles, they differ in that they contain different beneficial bacteria and yeasts. Kefir contains a much larger range of bacteria and yeasts than yogurt, and kefir has the ability to colonize the intestinal tract in a way that yogurt does not. Kefir is easily digested and provides a unique and intense cleanse to the intestinal tract. Kefir is a very nourishing and balanced food that strongly contributes to a healthy immune system. Kefir is also calming to the nervous system and may help combat depression, sleep disorders, and other nervous conditions.[5] Kefir is a potent tool for balancing the body's ecosystem; food as medicine indeed.

When making coconut kefir, there are two options. The first is using milk kefir grains, which are slightly trickier and more challenging to use, and the second is using a kefir starter culture in powdered form. The latter method is a little easier and less finicky. Both yield delicious and beneficial results.

It is very important to use fresh coconut cream. Boxed or canned coconut milk or cream with fillers and added sweeteners can disrupt the fermentation process and kill the grains. When making kefir, it is also very important that the grains do not touch any metals. Metal will kill the cultures. Use only wooden spoons and plastic strainers.

option one | **milk kefir grains**

1 packet of active milk kefir grains, or about 2 tablespoons of active culture*
2 cups coconut cream

*Milk kefir grains must be activated in pasteurized dairy milk in order to work. This contributes trace amounts of dairy to the final product but does not pose a lactose issue. Please see the resources section for a recommended source of high-quality kefir grains and starter cultures.

In a large jar, stir the kefir grains and the coconut cream together. Cover with cheesecloth or light gauze. Make sure that the mixture can breathe but will not get contaminated. Culture at room temperature for at least 12 hours. You can allow the mixture to ferment for up to 24 hours, depending on how tart you want it. Once the kefir has reached a desired flavor and consistency, using a non-metal strainer, strain the grains; the kefir is ready to serve.

Makes 2 cups.

The grains may not culture properly during the first use. If they don't, reuse the grains to start another batch. Make sure they are handled properly and have not been contaminated with metal or other harsh substances. You can continue to use active cultures in coconut cream, but they should be revitalized in dairy milk for twenty-four hours once a week.

option two | **kefir starter culture**

Using a kefir starter culture is much easier than handling kefir grains, but it must be noted that this culture is also not completely dairy-free as it contains dehydrated milk solids. The result is a safe choice for dairy-free diets or lactose intolerance, but it is not completely vegan. You can also use this culture in coconut water, juice, or any liquid that contains some sugars.

1 packet of powdered kefir starter culture
4 cups coconut cream or liquid of your choice*

*Make sure the liquid is room temperature (at least 68 degrees) before using.

In a large jar, stir the kefir starter culture and the coconut cream together. Cover with cheesecloth or light gauze, making sure the mixture can breathe but will not become contaminated. Culture at room temperature for at least 12 hours or overnight. Once the coconut cream is cultured, the mixture is ready to serve.

Makes 4 cups.

To "seed" or start another batch, use 1/4 cup fermented coconut cream as a base mixed with up to 3 3/4 cups liquid of your choice. This can be done three to four times before a new package is needed.

coconut cream kefir recipes

There are many variations and flavor combinations that are amazing with kefirs; feel free to experiment. The following combinations all have medicinal qualities and unique flavor profiles and can be made using either the grain or starter culture options.

coco rose

2 cups coconut cream kefir
1 tablespoon rose water
1 tablespoon white honey or coconut nectar
Pinch of cardamom
Pinch of sea salt

Blend and serve.

Makes 1 serving.

vanilla bean kefir

2 cups coconut cream kefir
Scrapings from 1 vanilla bean, or 1 teaspoon vanilla extract
2–3 fresh dates, pitted

Blend and serve.

Makes 1 serving.

blueberry kefir, cardamom, *and* white honey

This recipe is insanely delicious. When I first discovered the uniquely exotic flavor of cardamom, I wanted to use it in absolutely everything. It pairs brilliantly with a touch of sweetness, but it is delicious without it as well. Cardamom is commonly used in Indian cuisine, and it has also been recognized in Ayurvedic medicine as a treatment for ulcers, as helpful for digestive disorders, and perhaps able to uplift the spirits and fight off depression. It would certainly be hard to be in a bad mood when sipping this smoothie.

2 cups coconut cream kefir
1/4 cup blueberries
Pinch of cardamom
Pinch of sea salt
1 tablespoon white honey or coconut nectar, or 1 date (optional)

Blend all ingredients until smooth.

Makes 1–2 servings.

kefir ranch dressing

2 cups coconut cream kefir
1–2 tablespoons olive oil
1 teaspoon nutritional yeast
1 teaspoon miso
1 teaspoon fresh parsley
1 teaspoon basil
1–2 cloves of garlic
1/4 teaspoon black pepper
Pinch of cayenne
1 teaspoon minced onion (optional)
1/4 cup cucumber juice (optional)

Blend all ingredients until smooth.

Makes enough dressing for several salads.

coconut water kefir

Coconut water kefir is the only true non-dairy form of kefir (because water grains are used instead of dairy-based grains), but the health benefits are similar. Water kefir is lighter; to some, it is probably easier to drink and a little less intense than milk kefir. If you drink kombucha, beer, or wine, you'll notice similarities in the flavor profiles.

Water kefir grains feed off of the sugars in coconut water; they can also be used in any kind of juice or sugar water. The kefir can be fermented using dried fruit or fruit juice, or it can be flavored after plain fermentation. Water kefir is very versatile and easy to make once the grains have been activated. Once the grains are alive, you must continue to tend to them or they will die. As with all kefir, it is important not to contaminate the grains and to make sure none of the ingredients involved come in contact with any metal.

coconut water kefir

4 cups coconut water (raw is best, but pasteurized coconut water will work as well)
1/4 cup activated water kefir grains*

*To activate grains: Dissolve 1/4 cup coconut crystals in 3–4 cups hot water. Let water cool before adding the grains. Allow to sit at room temperature for 3–4 days until the grains become translucent and plump and look lively. Strain off sugar water using a non-metal strainer. The grains are now ready to be used immediately. Between fermentations, feed the grains with the strained sugar water mixture; this keeps them alive and thriving. The grains will multiply over time, so there will always be a fresh batch to brew.

Pour coconut water into a glass jar. Add activated grains and stir with a non-metal spatula. Cover jar to prevent contamination and allow mixture to brew at room temperature for 24–48 hours. Once kefir is fermented, pour through a plastic sieve to strain. Store kefir in a glass jar at room temperature or in the refrigerator to keep chilled. It is now ready to be flavored and served.

Makes 4 cups. Water kefir can last up to two to three weeks in the refrigerator.

○ *To make a more effervescent kefir, ferment it in a sealed jar or container to trap the CO_2 produced during the fermentation process. Bubbly kefir is delicious. This could be called "kefir champagne" or "kefir soda."*

Kefir grains soaking in coconut water

coconut water kefir recipes

blood orange soda

2 cups coconut water kefir
1/4 cup blood orange juice
2 tablespoons turmeric juice
A few drops of stevia (optional)

Stir all ingredients together.

Makes 1–2 servings.

apple, aloe, *and* mint

2 cups coconut water kefir
1/4 cup apple juice
2–3 tablespoons aloe vera
1/4 cup fresh mint leaves
A few drops of stevia (optional)

Stir all ingredients together. Allow mint to infuse for at least 30 minutes before serving.

Makes 1–2 servings.

kefir champagne

2 cups coconut water kefir, fermented under pressure
1/2 cup white grape juice
1 tablespoon lemon juice (optional)

Stir all ingredients together. Chill well before serving.

Makes 1–2 servings.

kefir cosmo

1 tablespoon cranberry juice extract
2 tablespoons orange juice
1–2 tablespoons coconut water kefir
Drop of stevia (optional)
Ice
Orange zest for garnish

Shake all ingredients with ice. Strain and serve. Garnish with orange zest.

Makes 1 serving.

kefir mojito

1 cup coconut water kefir
8–10 mint leaves, torn into pieces
1/2 cup fresh lime juice
2 tablespoons coconut sugar, or a few drops of stevia
Sparkling water
Ice
Extra mint and lime for garnish

Place ice in a cocktail shaker, then add the kefir, mint leaves, lime juice, and sugar or stevia. Shake well. Strain mixture and serve by dividing evenly between four highball glasses. Top each serving off with sparkling water. Garnish with mint and lime.

Makes 4 servings.

other coconut ferments

coconut sour cream

This recipe is the perfect substitute for traditional sour cream. It is virtually identical to the coconut yogurt recipe but contains more water and a splash of apple cider vinegar, lemon juice, and salt. If you don't want to make two separate recipes, add water and the remaining ingredients to the fermented yogurt recipe on the next page.

1 cup young coconut meat
1/2–3/4 cup water, depending on your preferred consistency
1/2 teaspoon probiotic powder, or 1 probiotic capsule
1 teaspoon apple cider vinegar
1 teaspoon lemon juice
1/2 teaspoon sea salt, or to taste

Blend coconut meat with water until smooth. Stir in probiotic powder. Place in a covered bowl in a dehydrator at 100 degrees for 8–10 hours, or ferment at room temperature for up to 12 hours. The sour cream needs to be in a warm place with good airflow.

Makes 1 cup.

coconut yogurt

Coconut yogurt is one of the easiest and most versatile ways to use the coconut. It is perfect as a substitution in any recipe that calls for plain, full-fat, Greek-style yogurt. Once fermented, the coconut imparts a light sweetness that isn't over-powering. This recipe will create a yogurt base that will be used other recipes throughout the book.

2 cups young coconut meat*
1/2 teaspoon probiotic powder
Water, if needed to blend

*Using Exotic Superfoods® coconut meat or another brand of prepared young coconut meat makes this recipe very easy and foolproof. If you choose to use fresh young coconuts, make sure that the meat is pure white and that it is of the best quality. Sometimes it can take up to four coconuts to yield two cups of coconut meat.

Blend coconut meat until smooth. Stir in probiotic powder. Place in a covered bowl in a dehydrator at 100 degrees for 8–10 hours, or ferment at room tempera-ture for up to 12 hours. The yogurt needs to be in a warm place with good airflow. Depending on the quality and strength of your chosen probiotic, the fermenta-tion time may vary. Look for a creamy, tangy flavor. If there is any discoloration or mold, discard the yogurt. If using clean equipment and containers, this should not be a problem. Once fermented, stir well. If fermented in the dehydrator, the yogurt will develop a thin film on top, which is still edible. This film can be dis-carded, but it does incorporate well into the yogurt.

Makes 2–4 servings. Coconut yogurt lasts in the refrigerator for up to one week.

○ *Make sure to flavor the coconut yogurt only once it is fermented. If ingredients are added before fermentation, the fermentation process will be altered and disrupted.*

VARIATIONS

Here are some simple combinations to enjoy. Stir the following ingredients into coconut yogurt for variation.

- **Honey lemon**—2 tablespoons lemon juice, 1 tablespoon honey
- **Mixed berry**—1/4 cup berries of your choice mashed with 1 tablespoon lemon juice and 1 tablespoon agave or honey
- **Raita**—2 teaspoons chopped dill, 1/2 cup diced cucumbers, 1 table-spoon lemon juice, pinch of sea salt, 1 teaspoon optional crushed garlic

Golden Coconut Milk (page 89)

breakfast

"Do you know what breakfast cereal is made of? It's made of all those little curly wooden shavings you find in pencil sharpeners!"

—ROALD DAHL

Breakfast is usually considered the most important meal of the day, but so often it is either nonexistent or far from healthy. Breakfast should not be used as an opportunity to fuel up on sugar and caffeine.

Accosting your body with food when you first wake up can also be a mistake. Waking up and eating a heavy meal does not always equal energy, especially if that heavy meal is full of refined carbs and sugars. If you choose to wake up and eat this way, you will probably crash and burn around lunchtime. Breakfast can easily become your worst meal of the day nutritionally.

Fortunately, breakfast can also be the healthiest meal, as it is often the meal that we have the most control over. Between midnight and midday, your liver works the hardest to eliminate toxins, so it is important to eat foods in the morning that support this process. Consuming juices, fruits, and citrus is a great way to replenish your cells and alkalize toxins from the night before without putting your liver to work. An acidic meal on top of an acidic meal is a recipe for distress.

Morning is an important time to balance your mind and body by establishing healthy and grounding rituals that support your day. Frenetic mornings equal frenetic minds and actions. Before consuming any of the recipes below, try drinking a room-temperature glass of lemon water. Lemon water is a grounding addition to a morning routine. Warm water with lemon is a natural immunity booster, metabolism booster, alkalizer, powerful lymph hydrator, and digestive aid. Lemon water in combination with a coconut-centric breakfast is *the* recipe to get the glow. Clearer eyes, clearer skin, and a clearer mind are all added benefits.

Eat breakfast, but make it healthy. This chapter is full of recipes that support that process.

strawberries *and* cream breakfast quinoa

This recipe is very versatile and can be made using any combination of fruit and spices, or you can keep it simple and leave the spices out. Peaches, plums, pears, and figs are all excellent in this dish. Because the prep is a little more time-consuming, this makes a great recipe for the weekend—and as leftovers, it is divine.

2 tablespoons coconut oil
1 cup quinoa, rinsed
1 cup coconut milk
Pinch of sea salt
1 cup water
1/2 cup strawberries, diced
1/4 teaspoon cinnamon
2 tablespoons rose water (optional)
1 tablespoon honey or coconut nectar (optional)

Heat oil in a medium saucepan over medium heat. Add quinoa and cook, stirring often, until golden (about 5 minutes). Add coconut milk, salt, and water and stir to combine. Bring to a boil; reduce heat, cover, and simmer until quinoa is tender and liquid is evaporated, approximately 20 minutes. Stir in strawberries, spices, and rose water while there is still a little liquid left in the quinoa. Cook for about 5 more minutes until strawberries are soft. Let sit for 10 minutes before serving. Fluff with a fork.

Makes 2–4 servings.

buckwheat *with* mulberry *and* coconut granola

Buckwheat is great for digestion. It is a warming food, so it's great to consume in the winter months. It is also an excellent source of protein. Buckwheat is not technically a grain, so it is safe for people with celiac disease or for anyone that wants to avoid gluten. If you can't find dried mulberries, substitute raisins, chopped dried figs, or other dried fruit.

1 1/2 cups sprouted, dehydrated buckwheat
1 apple or pear, diced or roughly chopped
1/2 cup coconut flakes
1/2 cup golden raisins
1/2 cup dried mulberries
3/4 cup coconut nectar or date paste*
1 tablespoon coconut oil
1 teaspoon lemon or lime zest
1 teaspoon vanilla extract
1/2 teaspoon cinnamon
Pinch of sea salt
Pinch of cardamom (optional)

To make date paste: blend 5–8 soft dates with 1/4–1/2 cup water until a paste is formed.

Mix all ingredients by hand in a large bowl until well combined. Sprinkle onto a dehydrator sheet, allowing space between granola pieces. Dehydrate at 115 degrees for 18–24 hours or until dry, or bake at your oven's lowest temperature for 2–4 hours.

Makes about 1 quart.

mango lassi *with* turmeric, rose, *and* cardamom

A lassi is a traditional Indian yogurt-based beverage. Lassis are a blend of yogurt, water, spices, and sometimes fruit. Everything in this recipe is as close to the original as possible except for the substitution of fresh coconut yogurt and optional added sweetener. The flavor of this drink is exotic and intoxicating. If you have this for breakfast, you will set quite the tone for the rest of your day.

1/2 cup coconut yogurt
2 cups coconut water
1/2 cup diced mango, fresh or frozen
1 teaspoon turmeric powder, or 1 tablespoon turmeric juice
Pinch of cardamom
Pinch of sea salt
1 tablespoon raw honey or coconut nectar (optional)
1 tablespoon Royal Sense® Bulgarian rose water (optional)

Combine all ingredients in a blender until smooth and creamy.

Makes 1–2 servings.

> *Royal Sense® Bulgarian rose water (rose hydrosol) has enormous health and beauty benefits. Bulgarian roses are grown in the Rose Valley in Bulgaria where the roses are considered to be some of the most medicinal and beneficial in the world. Taken internally, rose water promotes healthy digestion, works as a detoxicant, balances energy and mood levels, and creates an overall sense of well-being. It is a cleansing and nourishing addition to a morning routine.*

peaches *and* cream parfait

Soaking or culturing grains is an ancient process that helps break down anti-nutrients and release all of the beneficial nutrients from the grain. Grains and legumes contain phytic acid that binds to important minerals and prevents your body from absorbing them; basically, it is a nutrition blocker. Soaking and souring breaks down the phytic acid and increases the nutrient availability. Soaking grains is also a great technique to use if you are gluten intolerant (even if you are using gluten-free grains).

2 peaches, peeled and diced (approximately 2 cups)
1 cup coconut milk
1 tablespoon honey or coconut nectar (optional)
1/2 teaspoon cinnamon
1 teaspoon vanilla extract
Pinch of sea salt
1/2 cup chopped walnuts
1 cup soaked grains

Blend 1 cup of the diced peaches with coconut milk, honey, cinnamon, vanilla, and sea salt. Stir in remaining diced peaches and chopped walnuts. In a wide-mouth glass or glass canning jar, layer half of the peaches-and-coconut-cream blend with half of the soaked grains. Garnish with the ingredients of your choice.

Makes 2 servings. Soaked grains can be stored in the refrigerator for one day but are edible immediately after soaking.

SOAKING GRAINS

Soaking requires an acidic medium. This can include dairy-based acid mediums such as yogurt, whey, buttermilk, and kefir or non-dairy mediums like lemon juice, raw apple cider vinegar, coconut milk kefir, and water kefir. This recipe uses a non-dairy medium. Lemon juice is the preferred medium in this method; kefir works well, but if you don't have it on hand, lemon juice and apple cider vinegar are great substitutions. This method works well on most whole grains. Simply cover the grains with warm water and add approximately two teaspoons of the acidic medium. Soaking overnight or for twelve hours is enough time to break down most grains. Buckwheat groats have the shortest soaking time and are ready in about seven hours.

The one exception to this formula is oats. Oats contain a large amount of hard-to-digest phytates, and soaking them is not necessarily enough to properly break them down. To do so, add a small amount of a phytate-rich grain such as rolled rye oats or buckwheat groats to the mixture. One cup of oats, two teaspoons of acidic medium, and one tablespoon of a phytate-rich grain with a twenty-four-hour soak time is enough to make oats properly digestible. Select the grain of your choice for this recipe.

coconut meat crudités
with coconut coffee dip

Think of this dish as breakfast crudités. Mature coconut is excellent on its own, but this is a fun variation to make with leftover coffee grinds. This dip lasts for several weeks in the refrigerator and serves four to six people, depending on how hungry you are. Substitute maca (unless pregnant or nursing) or a medicinal mushroom powder like reishi for variation.

1 cup coconut nectar, honey, or agave
2 tablespoons coconut oil
2 tablespoons water
1 teaspoon vanilla extract
1 teaspoon coffee grinds, or a pinch of sea salt
1/4 teaspoon cinnamon
1 tablespoon coconut butter (optional)
1 teaspoon maca (optional)

Blend all dip ingredients until smooth. Serve with 1 cup mature coconut slices.

Makes 6–8 servings.

golden coconut milk

The term "golden milk" has become synonymous with turmeric-enhanced milk. Golden milk is one of the most delicious smoothie and nut milk variations. The combination of coconut milk, honey, and fresh turmeric is a powerful cold and flu fighter, immunity booster, and mood enhancer. I've included the optional royal jelly version, which takes this recipe to the next level of nutrition. This recipe is great on its own or when served in a yogi tea blend.

2 cups coconut milk
1 tablespoon honey
1 tablespoon fresh turmeric juice, or 1 teaspoon dried turmeric
Pinch of cinnamon
Pinch of nutmeg
1 serving of Immortality Alchemy® Bee Mana (optional)

Blend all ingredients until smooth.

Makes 1–2 servings.

Immortality Alchemy® Bee Mana is 100 percent raw, organic, dried royal jelly extracted from bees that are never fed sugar or artificial sweeteners. Royal jelly is a rich source of essential vitamins (including B12, B5, and folic acid), minerals, proteins, and amino acids. It can be used as a general health tonic for increasing energy and stamina and boosting immunity.[6]

coconut yogurt *with* stone fruit, sprouted buckwheat, *and* lemon zest

There is something so satisfying about yogurt for breakfast. Not only is the morning probiotic boost a healthy way to start the day, but there is also a comforting and familiar coziness about eating yogurt. Unfortunately, most yogurt on the market—including the non-dairy versions—is full of synthetic fillers and sugars. Making a base recipe of yogurt on a Sunday evening is a great way to set a healthy tone for the rest of the week. One recipe will last about five days. This is just one of many examples of how to get creative with your morning breakfast bowl.

1/2 cup coconut yogurt
1 medium plum, peach, or other stone fruit, thinly sliced
1 tablespoon lemon zest
1 tablespoon raw honey or coconut nectar (optional)
Pinch of sea salt (optional)

Arrange each yogurt bowl as you wish with the ingredients listed above. The sweetener and lemon zest can be added as garnishes or stirred into the yogurt.

Makes 1–4 servings, depending on how much yogurt you have on hand.

- *Never flavor yogurt before fermenting; only flavor after fermentation is complete. It's best to flavor one batch at a time.*

- *Any seasonal fruit works in this recipe. When stone fruit is available, it's one of the best choices. Citrus fruits like grapefruits or oranges are also excellent choices in this dish. The addition of raw honey or coconut nectar is optional, but it brings a touch of sweetness to balance out the tart.*

golden coconut chia pudding

Chia is a great source of protein and a balanced source of omega fatty acids. Chia seeds contain omega-3, which is often lacking in our diets, especially if fish has been eliminated. When the omega fatty acids in our diets are out of balance, many common ailments occur, such as dry skin, dandruff, moodiness, and irritability. Chia is also a strong anti-inflammatory agent. In combination with turmeric, this is a powerfully medicinal concoction full of inflammation-fighting compounds, antioxidants, and protein. You could definitely call this a superfood cereal.

2 cups coconut milk
1/4 cup coconut nectar
1 teaspoon turmeric
1 teaspoon cardamom
1 teaspoon cinnamon
1 teaspoon freshly grated ginger
1 teaspoon vanilla extract
Pinch of sea salt
1/4 cup raisins
2/3 cup chia seeds
1 tablespoon honey, or 1 teaspoon Immortality Alchemy® Bee Mana
    (optional)

Combine coconut milk with all ingredients except raisins and chia seeds and blend until smooth. Stir in raisins and chia seeds. Allow mixture to sit for at least 2 hours before serving. It is best if it is allowed to sit in the refrigerator overnight. Before serving, garnish with raisins, coconut flakes, and a pinch of cinnamon.

Makes 6–8 servings.

coconut yogurt *and* carrot ginger cream smoothie

It took me a long time to like carrot juice in any form. I do have fond memories of seeing women with their carrot-juice mustaches in the original juice bars. Although the mustache is no longer in vogue, carrot juice is still around. Now, I absolutely love it and think it is brilliant with the flavor of coconut. Carrots are a sweet vegetable with a high sugar content, but they are a great way to get a boost of vitamin A, vitamin C, potassium, and other macronutrients into your diet. Carrots contribute to skin health, lung health, and proper digestion and boast a wide range of other benefits. The flavor of this smoothie is actually a bit like pumpkin pie.

1/2 cup coconut yogurt
1 cup coconut water
1 cup carrot juice
Pinch of nutmeg
Pinch of cinnamon
Pinch of sea salt
1 tablespoon ginger juice (optional)

Blend all ingredients until smooth.

Makes 1 serving.

coconut matcha smoothie

A cup of matcha tea is one of the best things on Earth. Matcha is one of nature's most powerful detoxifiers, energizers, and antioxidants. Matcha helps focus energy—unlike coffee, which sometimes causes irritability and nervousness. Matcha is equally awesome when added to smoothies, granola, ice cream, and nut milk. It has a slight vegetable flavor with a sweet and creamy aftertaste.

1/2 cup young coconut meat
1 1/2–2 cups coconut water
1 teaspoon matcha powder
Pinch of sea salt
1 teaspoon honey or coconut nectar (optional)

Blend all ingredients until smooth.

Makes 1 serving.

○ *Make sure that you are using pure, high-quality matcha powder. There are products on the market that are sweetened with poor-quality sweeteners and fillers and are far from pure. It is really important to make the investment and buy quality matcha—a little goes a long way.*

Coconut with Spinach, Golden Raisins, and Pine Nuts (page 115)

lunch

"Ask not what you can do for your country. Ask what's for lunch."

—ORSON WELLES

Lunch is one of the best times of day to consume coconut. Coconut has such fueling powers—consuming it midday will leave you feeling satisfied, grounded, and energized without any afternoon sluggishness. Most of us have no time to eat a healthy lunch, much less prepare one; each of these recipes is crafted with that in mind. All of the more time-consuming recipes in this chapter can be made at the beginning of the week and can be used all week long. They key to eating well is to be prepared.

Dressings, coconut bacon, coconut mayonnaise, and hummus are all amazing condiments to have on hand. Even if you are eating lunch in front of your computer screen, make sure to chew thoughtfully and be as present as possible. Lunch can easily be a very nourishing and grounding point in the day.

cblt *(Coconut Bacon, Avocado, Romaine, and Tomato)*

This recipe is a fresh take on the BLT. There is something so satisfying about crisp romaine lettuce in combination with creamy mayonnaise, avocado, sweet tomatoes, and smoky coconut "bacon." It is the quintessential summer fare. If you don't feel like making the coconut bacon, this dish is also delicious with cured olives or baked mushrooms. However, I like to keep coconut bacon on hand because it is good with just about everything.

COCONUT MAYONNAISE

1 cup young coconut meat
1/4 cup coconut oil
1/4 cup olive oil
2 teaspoons Dijon mustard
1 tablespoon lemon juice
Generous pinch of sea salt
Freshly ground pepper

Blend all ingredients until smooth. Keeps well in the refrigerator for three to four days.

Makes 2–4 servings.

COCONUT BACON

1 small deseeded chipotle chili
1/4 cup coconut nectar
1 tablespoon cumin
1/2 cup water
1 tablespoon coconut oil
1 tablespoon coconut aminos
1 1/2 tablespoons coconut vinegar or apple cider vinegar
1 teaspoon sea salt
1 teaspoon smoked paprika
1 tablespoon mesquite powder
Pinch of cayenne, or more if you like a spicier flavor
3 cups large coconut flakes

Combine all ingredients except coconut flakes in a blender or food processor until smooth. Stir in coconut flakes. Dehydrate overnight 8–10 hours or until completely crispy, or bake at your oven's lowest temperature for 2–3 hours. Coconut bacon can be stored in an airtight container for up to one month.

FINAL ASSEMBLY
3 romaine lettuce leaves
1 tablespoon coconut mayonnaise per serving
Approximately 1/4 cup coconut bacon per serving
1 small or medium heirloom tomato, chopped
1/2 avocado, diced
A few teaspoons diced Persian cucumber
A few teaspoons chopped parsley

Fill each romaine leaf with a generous portion of coconut mayonnaise. Then fill leaves with coconut bacon, tomato, avocado, cucumber, and parsley.

Makes 2–3 CBLT servings.

curried kale *with* carrots, currants, *and* cashews

This salad is a variation of a classic massaged kale salad and is excellent when served with brown rice or quinoa. Feel free to add or subtract ingredients as you would like.

SALAD
1 large head of kale
1 teaspoon lemon juice
1 teaspoon coconut oil (optional)
Pinch of sea salt

Remove kale stems and tear leaves into large pieces. Massage kale with lemon, coconut oil, and a pinch of salt to break down the fibers.

DRESSING
1/2 cup coconut butter
1–2 tablespoons water
1 tablespoon rice vinegar
1 tablespoon coconut aminos or tamari
1 tablespoon lemon juice
1 1/2 teaspoons red curry paste
1/4 teaspoon turmeric
Pinch of cayenne
Generous pinch of sea salt
1–2 dates (optional)
2 tablespoons minced cilantro

Blend all ingredients except cilantro until smooth and fully emulsified. Stir in cilantro.

FINAL ASSEMBLY
1 cup shredded carrots
1/4 cup currants
1/4 cup cashews, chopped
Cilantro garnish (optional)

Toss kale with approximately 1/3 of the dressing. Add carrots, currants, and cashews; dress everything well, but be careful to not wilt the salad. Garnish with a few chopped cashews and cilantro.

Makes 2–4 servings.

coconut caesar *with* romaine, dulse, *and* cured olives

This salad can be served as a romaine "wedge" inspired by restaurant platings. You can also use romaine hearts or any other crisp lettuce. Oil-cured or sun-dried olives are best, but you can dehydrate pitted olives to get this same effect. The concentrated salty flavor of dulse seaweed makes an excellent substitution for the anchovies that are traditionally found in Caesar salad. Dulse is rich in minerals and is an excellent source of protein. The dressing is great on its own and also makes a delicious dip for crudités.

COCONUT CAESAR DRESSING

1/2 cup young coconut meat
1/3 cup lemon juice
1 clove garlic
2 tablespoons white miso
1/2 teaspoon black pepper
1 teaspoon dulse flakes or finely chopped dulse
1 teaspoon apple cider vinegar
1–2 tablespoons water, or more as needed
1/2 tablespoon nutritional yeast (optional)
1–2 tablespoons olive oil (optional)
1/4 cup pine nuts, or 1/4 avocado (optional)*

The addition of either pine nuts or avocado will make a thicker, creamier dressing.

Blend all dressing ingredients together until smooth.

SALAD

2–3 small heads or 1 large head of romaine lettuce
1/2 cup chopped dulse or dulse flakes
1/4 cup sun-dried or oil-cured kalamata olives, pitted and chopped

Slice romaine hearts on the bias. In a large bowl, gently toss lettuce with the coconut dressing, making sure to fully coat leaves for a wedge-style salad. Garnish with chopped olives and dulse. Try tearing leaves and tossing all ingredients together for a more traditional Caesar salad presentation.

Makes 2–4 servings.

sprouted mung bean salad *with* amaranth, avocado, *and* curry

This salad is packed with super nutrients. Sprouted mung beans are crisp and nutty in flavor and provide an excellent source of protein and carbohydrates. Amaranth leaves and sprouts are also highly packed with vitamins and minerals and provide a powerful boost of energy to the body. If you can't find amaranth, substitute spinach or another leafy green. The dressing on its own is a beauty elixir with potent anti-inflammatory and detoxifying properties.

DRESSING

2–4 tablespoons coconut oil
1 teaspoon coconut nectar
2 tablespoons lemon juice, or juice of 1 lemon
1/4 teaspoon yellow curry powder
Generous pinch of sea salt
Freshly ground pepper, to taste
1/4 teaspoon chia seeds (optional)

Blend or whisk all ingredients together until smooth and emulsified.

SALAD

2 cups sprouted mung beans (purchase these or sprout your own)
1 cup shredded beets
1 cup amaranth leaves or amaranth sprouts
1 avocado, sliced
Edible flowers (optional garnish)

Toss mung beans and beets with 1/3 cup dressing to fully coat. If possible, allow mixture to marinate 10–15 minutes before serving. Gently toss in amaranth. To serve, garnish with avocado, edible flowers, and a drizzle of dressing.

Makes 2–4 servings.

hijiki *and* coconut seaweed salad

Hijiki, like most seaweed, is full of health and beauty benefits. Hijiki is a brown seaweed that has been a part of a balanced diet in Japan for centuries. Hijiki is rich in dietary fiber and minerals such as calcium, iron, and magnesium. Marine plants in all forms have the ability to revitalize skin, hair, and overall health— they are nature's beauty solution. Culinarily speaking, hijiki has a slightly meaty texture that is very satisfying. The shredded coconut adds unexpected sweetness that cuts through the seaweed's salty flavor and balances out this dish.

1/2 cup dried hijiki
1/4 cup dried coconut flakes
1/4 cup deseeded and diced cucumber
1/4 cup minced cilantro leaves
2 tablespoons lemon juice
1 tablespoon coconut aminos, tamari, or soy sauce
1/2 tablespoon sesame oil
1/4 teaspoon sea salt
Pinch of cayenne
1/2 avocado, diced
1/2 tablespoon sesame seeds (optional)

Cover hijiki with cold water and soak for 10 minutes. Drain hijiki well and press out any extra liquid by hand. Mix in all remaining ingredients except for avocado and sesame seeds and allow the mixture to marinate for at least 10 minutes. Before serving, add avocado. Garnish with sesame seeds.

Makes 2–4 servings.

○ *As long as you make sure to add the avocado last, you can make this recipe several days in advance. It will last up to four days in the refrigerator.*

sea palm *with* young coconut, carrot, *and* miso

Sea palm is a brown kelp seaweed unique to the western coast of North America. Sea palm is very hearty and is one of the few types of seaweed that can survive out of water. It thrives on rocky coasts with constant waves. The texture is very unique and it makes a great addition to salads, grain bowls, or virtually any vegetable dish. Sea palm isn't necessarily for the faint of heart; if someone doesn't like seaweed, they may not love this dish. Sea palm is relatively rare and can be difficult to find, so if you aren't able to find a good source, wakame and hijiki both make great substitutions. Like all seaweed, sea palm is an amazing source of vitamins and trace minerals. The combination of coconut and seaweed is natural magic—this salad is a superfood.

1 1-ounce package of sea palm fronds, rinsed well and drained
2 tablespoons white miso
2 tablespoons lemon juice
1–2 tablespoons water
1/2 cup shredded carrot
1 cup thinly sliced young coconut meat

Soak sea palm fronds in a large bowl of warm water for 30 minutes or until softened. The seaweed should have a gelatinous, al dente, noodle-like texture. Drain and rinse well. Whisk together miso, lemon juice, and water in a small bowl. Add carrot, sea palm fronds, and coconut and toss to coat.

Makes 2–4 servings.

manna bread *with* radish, mint, *and* coconut mayonnaise

Open-faced radish sandwiches are the perfect excuse to use coconut mayonnaise again. Use any bread that you like, but a sprouted or gluten-free bread option is preferred.

2–4 slices of bread
1/4 cup coconut mayonnaise (see page 100)
Several pinches of sea salt (a flaky salt like Maldon® works best)
1/2 cup sliced radishes, optionally tossed with a little lemon juice and
    olive oil
1 avocado, diced
Mint or basil microgreens

Spread a few tablespoons of mayonnaise on each bread slice. Add a few pinches of sea salt. Top with radishes, avocado, microgreens, and a few more pinches of sea salt.

Makes 2–4 servings.

coconut curry soup

This dish is part soup, part curried smoothie. This is a perfect on-the-go lunch—it's easy to digest and tastes amazing. The addition of carrot juice further boosts the nutrition, but if you don't have time to juice, filtered water or coconut water both taste great.

1/2 cup young coconut meat
1 1/2 cups coconut water
1 1/2 cups carrot juice, filtered water, or vegetable stock
1/2 tablespoon curry powder
1 tablespoon grated ginger
1 teaspoon grated lemongrass
1/4 cup diced tomatoes
Pinch of cayenne
2 tablespoons lime juice
1/2 tablespoon miso
Pinch of sea salt, or more to taste
1 tablespoon coconut oil (optional)
1 clove garlic (optional)
1 tablespoon minced onion (optional)
4 tablespoons cilantro leaves (optional garnish)
3–4 slices of lime (optional garnish)

Blend all ingredients until smooth and fully emulsified. Garnish before serving.

Makes 2–4 servings.

coconut curry hummus

It wouldn't be lunch without hummus. Try adding a tablespoon or two of hummus to serve as a creamy dressing on a salad mixed with some chopped avocado. This is also a great recipe for sandwiches, for wraps, as a dip, and so on. The flavors of hummus and coconut strike a perfectly unique balance.

1 cup coconut yogurt
2 tablespoons shredded, unsweetened coconut
1/2 cup chickpeas or peeled and chopped zucchini
2 tablespoons tahini
1/4 cup water (more or less as needed)
2 tablespoons lemon juice
1 teaspoon curry powder
1/2 teaspoon turmeric
1/2 teaspoon ground ginger
Pinch of cayenne pepper
Pinch of sea salt, to taste

Blend all ingredients until smooth.

Makes 4–6 servings.

coconut *with* spinach, golden raisins, *and* pine nuts

The execution of this recipe has been simplified so that you can throw it together before you head out the door; by the time lunch comes around, the flavors will be perfectly mixed. The addition of coconut makes this more fulfilling than a simple salad. Try adding a generous amount of freshly ground pepper and an avocado.

2–3 large handfuls of spinach
2 tablespoons shredded coconut
2 tablespoons golden raisins
1 tablespoon pine nuts
A few splashes of apple cider vinegar
A few splashes of olive oil
Pinch of sea salt
Pepper, to taste

Toss all salad ingredients together with a few splashes of apple cider vinegar, olive oil, and a pinch of sea salt.

Makes 1 serving.

Sweet Potato with Coconut and Broccoli (page 120)

dinner

"It seems to me that our three basic needs, for food and security and love, are so mixed and mingled and entwined that we cannot straightly think of one without the others. So it happens that when I write of hunger, I am really writing about love and the hunger for it, and warmth and the love of it and the hunger for it . . . and then the warmth and richness and fine reality of hunger satisfied . . . and it is all one."

—MFK FISHER

Dinner can be considered the most important meal of the day, not because we should outrageously fill ourselves before we go to bed, or because eating it will make our waistlines the most trim, but because it has the potential to be the most spiritually satisfying and uplifting meal. When we eat dinner, we have the opportunity to connect with friends and loved ones and to digest a meal in a way that other times of the day don't allow. Eating a nice dinner gives us the opportunity to connect. This connection makes our digestion better and makes the nutrients in our food more available. I believe that having a thoughtful meal in the evening is an important recipe for building health and wellness.

Coconut is a powerful tool in preparing a beautiful end-of-the-day meal. Most of these recipes are centered around using coconut meat to create new and unique dishes in mostly raw food preparation, but it can be much simpler than that. Coconut oil is a great substitution to use in the kitchen as an oil replacement in all cooking and baking.

Heat, along with light and oxygen, can destroy the beneficial fats in most oils. But pure coconut oil, mainly a saturated fat, is able to withstand higher cooking temperatures. This chapter represents a selection of both raw food and cooked dishes. Start incorporating and experimenting with coconut oil in all of your cooking and baking needs. Simple dishes like spinach sautéed with coconut oil and a few chili flakes is incredibly grounding and satisfying.

sweet potato *with* coconut *and* broccoli

This dish is part bowl, part curry, and part hippie culinary experience. Try keeping the cilantro leaves whole. If you aren't a cilantro lover, feel free to substitute spinach, basil, and parsley and make your own variation. It's also amazing drizzled with tahini (and makes great leftovers).

2 large heads of broccoli, cut into bite-sized florets
2 large sweet potatoes, peeled, if desired, and cut into 1-inch cubes
2–3 tablespoons coconut oil
1 cup cilantro leaves
1–2 tablespoons coconut butter
1 avocado, diced
1 tablespoon curry powder
Juice of 2 limes
1 teaspoon red chili flakes
1/4 teaspoon sea salt

Steam or blanch the broccoli florets for 10 minutes maximum. Make sure the broccoli is bright green and crisp. Preheat oven to 400 degrees while prepping the sweet potatoes. Toss the sweet potato cubes in 1 tablespoon of the coconut oil with a few pinches of sea salt and an optional dash of cayenne pepper. Bake for 30–40 minutes until the sweet potato cubes are easily pierced with a fork, but not mushy. Once the broccoli and the sweet potatoes are cooked, toss them with all remaining ingredients. The avocado will become soft and create a creamy dressing. Season with as much salt, lime, and chili as you would like. Serve in large bowls.

Makes 2–4 servings.

parsnip noodles *with* coconut alfredo

This nut-free alfredo sauce is great with any type of noodle and can be served with your favorite cooked pasta or with your choice of kelp noodles.

NOODLES
4 large parsnips
Pinch of sea salt
Pinch of lemon juice

Wash and peel parsnips. Use a spiral slicer or mandoline to create long, thin strips of parsnip. Toss parsnip noodles with a few pinches of salt and lemon juice to tenderize them. Allow noodles to marinate while you make the alfredo.

ALFREDO SAUCE
1 cup young coconut meat
1 tablespoon olive oil
1/2 cup filtered water
1/4 teaspoon black pepper
2 tablespoons lemon juice
2 tablespoons apple cider vinegar
1 tablespoon nutritional yeast
1 tablespoon minced basil
1/2 teaspoon nutmeg
1/4 teaspoon sea salt
1 small clove of garlic (optional)

Blend all ingredients until smooth. Salt and pepper to taste.

FINAL ASSEMBLY
Pine nuts (optional garnish)
Basil (optional garnish)
Nasturtium leaves (optional garnish)

Strain any water off of the parsnip noodles and massage them with your hands until soft. Toss parsnip noodles with coconut alfredo sauce. Allow mixture to marinate at least 10 minutes before serving. Plate individually or serve family style. Garnish with pine nuts, basil, and nasturtium leaves if desired.

Makes 2–4 servings.

coconut ceviche *with* tomatillo, lime, *and* cilantro

This is one of the easiest and most impressive dishes in the book. Including the avocado is optional. Try serving this with sprouted corn tortillas and a side of guacamole to make ceviche tacos. Finish with your favorite organic hot sauce and a glass of rosé.

2 cups young coconut meat, or 1 bag Exotic Superfoods® young coconut meat, thinly sliced into 2 inch–long strips

1/2 cucumber, deseeded and diced

1 cup quartered tomatillos, or 1 cup quartered cherry tomatoes

1/2 cup torn cilantro leaves

1/2 jalapeño, deseeded and diced (use more or less depending on your heat tolerance)

Juice of 2 limes

1/4 cup coconut water

1 tablespoon olive oil

Sea salt, to taste

1 avocado, diced (optional)

Stir together all ingredients. Allow mixture to marinate for at least 30 minutes.

Makes 2–4 servings.

O   *This dish makes great leftovers and is actually better the next day.*

kelp curry *with* shiitakes, zucchini, *and* cashews

The noodles in this recipe are made from kelp, an edible brown seaweed that contains high amounts of iodine. Kelp noodles are gluten-free and virtually calorie-free and make a really easy pasta alternative. On their own, they don't have much flavor, so they function best as a carrier for different sauces. Feel free to experiment with different vegetables and combinations.

NOODLES

2 bags kelp noodles

2–3 teaspoons baking soda

1 tablespoon coconut aminos or tamari

1 tablespoon coconut nectar

2 tablespoons sesame oil

Rinse kelp noodles thoroughly in warm to hot water. Add baking soda and allow noodles to sit for 15 minutes. Rinse and drain well. Toss with all remaining ingredients and allow noodles to marinate for an hour before serving. These can be prepped a day in advance.

○ *Adding baking soda is the trick to softening kelp noodles before serving. The baking soda cooks the kelp noodles and makes them less crunchy.*

CURRY

1 1/2 cups young coconut meat

2 tablespoons lime juice

1 tablespoon lemon juice

1/2 Thai chili, deseeded and chopped (or more depending on your heat tolerance)

1 tablespoon chopped green onion

2 tablespoons chopped ginger

2 teaspoons yellow curry powder

1/4 cup coconut water

1 teaspoon sea salt, or to taste

Pinch of turmeric

Pinch of cayenne

Blend all ingredients until smooth.

SHIITAKES

2 cups shiitakes, sliced
1 tablespoon sesame oil
1 tablespoon coconut aminos or tamari
Pinch of sea salt

Toss all ingredients together and allow to marinate for at least 30 minutes at room temperature or in a dehydrator.

ZUCCHINI

2 medium zucchini, diced into 1/2-inch pieces
1 tablespoon coconut, olive, or sesame oil
Pinch of sea salt

Toss all ingredients together and allow them to marinate for at least 30 minutes at room temperature or in a dehydrator.

FINAL ASSEMBLY

1/4 cup chopped cashews
1/2 cup mung bean sprouts
Micro cilantro

Toss kelp noodles with curry sauce, marinated shiitakes, and zucchini until well combined. If possible, allow the mixture to marinate for at least 30 minutes before serving, but this is not necessary. You can warm the dish or individual servings in the oven or in a dehydrator if you prefer. Plate noodles, making sure to distribute ingredients evenly. Garnish with chopped cashews, mung bean sprouts, and micro cilantro.

Makes 4–6 servings.

coconut cauliflower rice

This rice makes an impressive and delicious side as part of an Indian dinner or platter, but my favorite way to serve it is as part of a bowl with steamed greens. The mint chutney isn't necessary, but it definitely adds another layer of flavor and enhances the dish. Any type of chili or hot sauce is also a great addition.

RICE
2 cups peeled and roughly chopped jicama

1 cup cauliflower florets

1 cup mature coconut meat, shredded in a food processor

1/2 cup diced celery

1–2 tablespoons coconut oil

1 tablespoon fresh lime juice

1/2 tablespoon coconut nectar

1 teaspoon turmeric

1/2 teaspoon cardamom

1/2 teaspoon crushed cumin seed

1/4 teaspoon sea salt, or more to taste

1/4 cup chopped golden raisins

1/4 cup shredded almonds or chopped cashews

1/4 cup minced cilantro

Process jicama, cauliflower, and coconut meat in a food processor until well combined and with a rice-like texture. Stir in remaining ingredients. Adjust seasoning to taste. If you like spice, add a bit of diced fresh or dried red chili.

○ *If you don't want to use jicama, try making this recipe using three cups of cauliflower.*

MINT CHUTNEY
1 cup cilantro leaves

1 1/2 cups mint leaves

1 green chili pepper

1 tablespoon lemon juice

1 tablespoon coconut nectar

Sea salt, to taste

1 tablespoon coconut oil (optional)

Process all ingredients in a food processor until well combined but still chunky. Serve coconut rice in bowls with steamed greens and a side of mint chutney.

Makes 2–4 servings.

pumpkin seed *and* brazil nut tacos *with* coconut sour cream

This is a fairly straightforward recipe that has seen multiple iterations in the raw food world. Fermented sour cream upgrades this version by adding a healthy dose of probiotics. This can be a create-your-own experience and is great to serve to large groups.

TACO MEAT

1 cup pumpkin seeds, soaked 1 hour and drained
1 cup Brazil nuts, soaked 1 hour and drained
1/2 cup sun-dried tomatoes, soaked
1 tablespoon extra virgin olive oil
1 tablespoon coconut aminos or tamari
1 teaspoon paprika
1/2 teaspoon chili powder
1/2 teaspoon cumin
1 teaspoon apple cider vinegar
2 dates, or 1 tablespoon coconut nectar
Pinch of sea salt
1/4 cup cilantro, chopped

Mix all ingredients in a food processor until well combined but still chunky. Adjust seasoning to taste.

FINAL ASSEMBLY

2 romaine hearts
1 cup coconut sour cream (see page 76)
1–2 heirloom tomatoes, diced
1 cup guacamole or diced avocado
Micro cilantro or cilantro leaves

You can serve this dish family style with all ingredients plated separately. To plate individually, use 2–3 romaine leaves per serving. Fill with sour cream, taco meat, tomatoes, and guacamole and garnish with micro cilantro or fresh cilantro leaves.

Makes 4 servings.

coconut sushi hand roll

Hand-rolled sushi makes a great lunch or dinner—it's a straightforward and uncomplicated presentation. Young coconut meat is an excellent fish substitute. Rice is not used in the following version, but this dish can definitely be served with any grain you would like; black rice offers a really nice contrast to the white coconut, and quinoa is also a great choice. Serve with a little pickled ginger, wasabi, and soy sauce or coconut aminos. Get creative!

3 sheets of nori, cut in half
1 cup shredded or julienned carrots
1 cup shredded or julienned beets
1 cucumber, deseeded and julienned
1 cup sliced young coconut meat
1 avocado, thinly sliced
A few handfuls of microgreens
A sprinkle of sesame seeds
1 shiso leaf (optional)

Place the nori sheet on the palm of your hand, shiny side down, and put a thin layer of sprouts or rice on the left third of the sheet. Sprouts, shredded carrots, and beets are best used as the "rice" because they are the driest ingredients. Place cucumber, coconut meat, and avocado vertically across the middle of the sprouts, beets, and carrots. Add the shiso leaf and a sprinkle of sesame seeds. Flip up the bottom left corner of the nori sheet and begin folding into a cone shape. Keep rolling until the cone is formed. Put a drop of water at the bottom right corner to serve as glue and close tightly.

Makes 2–4 servings, or use as many ingredients as you would like to serve a large group.

○ *Try serving hand rolls with as many ingredients as possible and letting everyone make their own.*

"green" coconut quinoa

This recipe was inspired by the *arroz verde* dish found in Mexican cuisine. Rich and refined, the flavor is truly decadent while still feeling healthy. You can serve this dish as a meal or as a side dish. Try serving it with grilled portobello mushrooms. Feel free to use rice or another grain.

1/2 cup cilantro leaves, tightly packed
1 cup spinach leaves
1 poblano chili, deseeded and diced
1 1/4 cups water
1 tablespoon miso
1 1/4 cups coconut milk
Pinch of sea salt
1 tablespoon coconut or olive oil
2 tablespoons coconut butter
1 1/2 cups quinoa
1/4 cup finely minced onion
1 clove garlic, minced

Combine cilantro, spinach, chili, water, and miso in a blender or food processor and process until smooth. Add milk and salt and blend a bit more until well combined. In a medium saucepan, heat oil and coconut butter over medium heat. When the coconut butter is melted, add the quinoa and sauté, stirring about every 30 seconds for 3–4 minutes. Add the onion and garlic and cook for 1 minute, stirring constantly. Add the contents of the blender, stir well, turn the heat to high, and bring to a boil. Cover the pan, turn the heat to very low, and cook for 20 minutes. Stir the quinoa carefully to avoid crushing it, cover it, then cook another 5 minutes. Remove the pan from the heat and let the quinoa steam in the covered pot for about 5–10 minutes. Serve hot.

Makes 4–6 servings.

carrot coconut wrapper
with pistachio pea hummus
and jalapeño cilantro oil

Coconut wrappers have become a staple in advanced raw food preparation. The following recipe is as beautiful as it is delicious. Feel free to cut the wrappers in any size you would like. They make great sandwich wraps, and you could easily take this dish in an even simpler direction by filling the wrapper with hummus, sprouts, and lettuce and making a satisfying wrap sandwich.

WRAPPERS

4 cups chopped young coconut meat
Pinch of sea salt
1/4 cup carrot juice
1 tablespoon turmeric juice, or 1/2 teaspoon turmeric powder

Blend coconut meat and salt in a high-speed blender until very smooth. Add carrot juice and turmeric. Spread mixture in a thin, even layer over a dehydrator sheet and dehydrate for 4–5 hours. Once fully dehydrated, remove meat from dehydrator sheet and cut into nine even wrappers. If the wrappers are too dry and brittle to cut, rehydrate them with a moist towel. The wrappers should be pliable.

Makes approximately 36 wrappers sized 3 1/2 by 3 1/2 inches.

HUMMUS

2 cups peas
2 tablespoons pistachios
2 tablespoons tahini
Juice of 1 lemon
2 teaspoons cumin powder
Sea salt and pepper, to taste
Water, as needed to blend
1 tablespoon coconut oil or extra virgin olive oil (optional)

Process all ingredients in a food processor until smooth. Add water as needed to yield a smooth consistency. This hummus is delicious on its own or when served with crudités. *(Recipe continued on next page.)*

OIL
1 jalapeño pepper
1/4 cup cilantro leaves
1/4 cup lime juice
1/4 cup coconut nectar, or 2 dates
2 teaspoons apple cider vinegar
Sea salt, to taste
1 clove garlic (optional)
1/2 cup extra virgin olive oil, or 1/4 cup olive oil plus 1/4 cup coconut oil

Blend all ingredients except oils in a blender until smooth. Strain the mixture through a fine mesh sieve. Add strained mixture back to blender and drizzle in olive oil or olive and coconut oils until the dressing is smooth and fully emulsified.

FINAL ASSEMBLY
Micro cilantro
Edible flowers

Fill each wrapper with 1 tablespoon pistachio hummus mixture. Close each wrapper on its diagonal, making a small triangular pocket. To serve, drizzle with jalapeño cilantro oil. Garnish with micro cilantro and edible flowers.

Makes 6–8 servings.

carrot ribbons *with* coconut almond satay

Coconut almond satay can be kept as a staple in the kitchen at all times. If you don't serve this as a pasta dish, use the sauce as a dip or spread for crudités and sandwiches. Carrots make the best noodle choice because of their sweetness and hearty texture, but feel free to use rice noodles, kelp, or another gluten-free noodle of your choice.

CARROT RIBBONS

5–6 cups packed carrot ribbons, made using a spiral slicer or mandoline
1 cup shredded cabbage
Pinch of sea salt
Juice of 1 lime
1 cup edamame
1 cup thinly sliced young coconut meat (optional)

Toss carrot ribbons and cabbage with sea salt and lime juice. Allow ribbons to marinate for at least 10 minutes before serving. Combine with edamame and coconut meat.

COCONUT ALMOND SATAY

1/4 cup almond butter
1/4 cup young coconut meat
1/4 cup water, or more as needed to blend
2 tablespoons coconut aminos or tamari
1 teaspoon miso
1 tablespoon coconut nectar, maple syrup, or agave
1 tablespoon apple cider vinegar
2 teaspoons curry powder
1 teaspoon grated ginger
Pinch of cayenne
Sea salt to taste
1 teaspoon minced garlic (optional)
1 teaspoon minced onion (optional)

Blend all ingredients until smooth. Add water as needed. It is best to make the sauce at least 1 hour before serving so that the flavors can marinate.

FINAL ASSEMBLY
Torn cilantro leaves
Torn mint leaves
1/4 cup chopped almonds (optional)

Toss carrot ribbons, cabbage, coconut meat, and edamame with at least 3/4 of the sauce, depending on how much you prefer. Allow flavors to marinate a few minutes before serving. Garnish with cilantro, mint leaves, and chopped almonds.

Makes 2–4 servings. If there is sauce left over, it makes a great dressing and will keep in the refrigerator for up to three days.

dessert

"Life itself is the proper binge."

—JULIA CHILD

Natural sweeteners that help create dessert without using dairy or artificial sugars are a gift to have access to in the kitchen. Whether or not you eat exclusively dairy-free, raw, vegan, or gluten-free, dairy-free desserts are deceptively delicious. Dessert is often the breaking point for many who are trying to eliminate refined sugars, refined flours, and dairy products, but it doesn't have to be. Crafting beautiful and healthy desserts is an empowering process. And, to be honest, serving someone a salad doesn't have nearly the impact as serving a healthy coconut cream pie or coconut ice cream does.

The coconut is a key component to many of these decadent creations and must be respected for its many culinary uses, especially at the dessert counter. Dairy-free desserts would not be nearly as delicious without it.

Each of the recipes in this chapter serves as a great basic template for any creative direction you might like to take. The coconut has such diverse use in the pastry kitchen that this chapter could easily have included a hundred more recipes, but further variations will be left up to you. Experiment with adding different spices, fruits, and flavors to these delicious concoctions.

coconut macaroons
with candied orange

Coconut macaroons were one of the first and most ubiquitously successful gourmet raw food treats to be in markets. There is something to be said about the simplicity and delicious flavor of these little sweets. Although there are many variations on this recipe, the following has been taken in the direction of Italian amaretti cookies; the combination of sweet and bitter is so delicious. Feel free to add a little orange liqueur to boost the sweet orange flavor.

1/2 cup raw macadamia nuts
1/2 cup raw almonds
2 cups shredded coconut
3 tablespoons coconut oil, melted
2 tablespoons orange juice
2 tablespoons orange zest
1/3 cup maple syrup
2 tablespoons coconut nectar
Scrapings from 1 vanilla bean, or 1 teaspoon vanilla extract
Pinch of sea salt
Pinch of cinnamon (optional)
Extra orange zest for garnish

In a food processor, pulse macadamia nuts and almonds until finely chopped to create a powder. Add coconut and sea salt and blend again until all ingredients are finely chopped and thoroughly mixed. Stir in liquids, orange zest, and spices and process until thoroughly combined. Using a small ice cream scoop, put cookies onto a mesh dehydrator sheet. Garnish with orange zest. Dehydrate for 14–16 hours at about 115 degrees, or bake at your oven's lowest temperature for 2 hours.

Makes approximately 2 dozen cookies.

coconut whipped cream

What's not to like about whipped cream? There is something so indulgent and sexy about this food. Coconut whipped cream can be substituted in any of your usual whipped cream applications. Coconut nectar is preferred in this recipe, but using clear agave keeps the color much whiter.

2 cups young coconut meat
1 cup coconut milk
1/2 cup coconut butter, melted
1/2 cup coconut nectar or clear agave
1 teaspoon lemon juice
Scrapings from 1 vanilla bean, or 1 teaspoon vanilla extract
1/4 cup coconut oil, melted

Blend all ingredients except coconut oil until smooth. Add the coconut oil and blend until fully emulsified. Freeze for approximately 30 minutes before serving.

Makes approximately 3 cups.

coconut cream pie

It wouldn't be a coconut kitchen without coconut cream pie. This recipe is a purely decadent coconut overload—it tastes completely sinful but is still relatively light because it isn't as nut heavy as other vegan desserts of this nature. You can mix it up by adding a cup of fresh fruit of your choice to the cream filling and some sliced fruit between layers; pineapple and a little ginger make great additions.

CRUST
1 1/2 cups macadamia nuts
1/2 cup shredded coconut
Several generous pinches of sea salt
3 tablespoons coconut sugar, or 2–3 dates
1 tablespoon melted coconut oil
1 teaspoon vanilla extract

In a food processor, pulse macadamia nuts, shredded coconut, and salt until they become a coarse crumble. Add the coconut sugar, coconut oil, and vanilla; lightly pulse until all ingredients are well mixed but not sticky. Add a few more pinches of sea salt. The crust should stick together only when pressed between your fingers.

FILLING
3 cups young coconut meat
1 cup coconut nectar
1 teaspoon vanilla extract
1 tablespoon lemon juice
Generous pinch of sea salt
1/2 cup melted coconut oil

Blend all ingredients except coconut oil in a blender until smooth. Add coconut oil and blend until fully emulsified.

FINAL ASSEMBLY
Coconut whipped cream topping (optional)
Toasted coconut flakes (optional garnish)

Press crust into a 9-inch tart pan with a removable bottom. Pour in coconut cream filling. Let sit in the refrigerator for at least an hour to firm. Top with coconut whipped cream and garnish with toasted coconut flakes. Refrigerate for at least 30 minutes before serving.

Makes one 9-inch pie.

lemon coconut bars

When life gives you lemons, make lemon bars. This version is just as sweet and decadent as the original. Irish moss is not used a lot in this book, but it is very helpful in making dairy- and gluten-free desserts. It is used as a binder and a thickener to replace eggs and gelatin. Irish moss is a type of red algae that grows primarily in the Atlantic Ocean. The flavor is ocean-like and sometimes hard to tolerate; for this reason, I find it is best used in desserts that have citrus and sweetener.

CRUST
1 cup macadamia nuts
1 cup shredded coconut
2–3 tablespoons coconut sugar
Pinch of sea salt
1 tablespoon coconut oil

Pulse dry macadamia nuts in a food processor until the result has the consistency of a fine flour. Add shredded coconut, coconut sugar, and sea salt and process until well combined. Add coconut oil. Make sure the mixture is crumbly but holds together when pressed between your fingers. Optionally add extra sea salt so that the crust tastes similar to shortbread.

FILLING
1/2 cup dry Irish moss*
3/4 cup lemon juice
1 tablespoon lemon zest
3/4 cup coconut nectar or maple syrup
1 teaspoon turmeric (for color)
A few drops of stevia
1/4 cup coconut oil, melted

When you purchase Irish moss, it usually comes dry and sandy, so it is important to rinse and soak it very well before using it in any recipes.

Rinse Irish moss thoroughly to remove dirt and debris; some moss will be sandier than other moss depending on the source. In a medium or large container, soak moss in water for a minimum of 4 hours, up to 8 hours maximum. Do not soak the moss too long or it will get waterlogged; Irish moss expands as it soaks. Rinse and drain moss well before using. In a high-speed blender, blend the drained Irish moss with the lemon juice, lemon zest, coconut nectar,

and turmeric until well combined. Taste and add a few drops of stevia if more sweetness is desired. Add coconut oil last and blend until fully emulsified.

MERINGUE

1 cup young coconut meat
1/2 cup plus 2 tablespoons coconut butter
1 teaspoon vanilla extract
1/4 cup coconut nectar or maple syrup
Pinch of sea salt

Blend all ingredients until smooth.

FINAL ASSEMBLY

Line a baking pan measuring 9 x 9 inches with parchment. Firmly press down the crust until it holds together well. Pour in lemon curd filling. Refrigerate for at least 8 hours before serving. Once filling is firm, cover with meringue and refrigerate for a few more minutes before serving. Slice into bars of your size preference. They are very light, so the serving can be large. Garnish with shredded coconut or the coconut dust created by lightly pulsing coconut flakes in a spice grinder.

Makes approximately nine 3-inch bars.

coconut ice cream

Ice cream is the quintessential comfort food for many of us; the sweet, cool, creamy flavor is nostalgic and familiar. There aren't many other foods that have quite the same cachet. This recipe comes remarkably close to the original. Although this recipe yields vanilla ice cream, it makes a great base for adding different flavors. Stir in cacao nibs and add a few drops of peppermint extract to make peppermint chip ice cream. Additionally, coconut nectar makes a simple substitute for caramel.

1 cup young coconut meat
1/2 cup coconut flakes
1/2 cup coconut nectar or white honey
1 1/4 cups coconut water or coconut milk
1–2 tablespoons coconut oil
1 tablespoon coconut butter
Scrapings from 1–2 vanilla beans, or 1 tablespoon vanilla extract
Generous pinch of sea salt

Blend all ingredients until smooth. Pour into ice cream machine and process according to manufacturer's instructions. Drizzle with coconut nectar.

Makes 1 pint.

○   *Substituting coconut water for coconut milk will make a much lighter sorbet-style version.*

matcha ice cream

Not only is matcha an energizer, an antioxidant powerhouse, and beneficial to cellular and brain function, but it also has an amazing flavor that lends itself beautifully to desserts. Matcha ice cream rivals chocolate and vanilla in terms of its timeless appeal. The flavor of matcha is delicious mixed with a little citrus. Feel free to experiment with the flavor by adding orange zest.

1 1/2 cups young coconut meat
1/2 cup coconut nectar or white honey
1 1/4 cups coconut water or coconut milk
1–2 tablespoons coconut oil
1 tablespoon coconut butter
1 tablespoon matcha tea powder
Scrapings from 1–2 vanilla beans, or 1 tablespoon vanilla extract
Generous pinch of sea salt

Blend all ingredients until smooth. Pour into ice cream machine and process according to manufacturer's instructions.

Makes 1 pint.

COCONUT ICE CREAM TOPPINGS

What's better than coconut ice cream? Coconut ice cream toppings. Below are three simple topping recipes that will enhance any ice cream recipe. Each recipe yields enough topping for several ice cream servings.

CHOCOLATE "MAGIC SHELL"

3/4 cup coconut oil
1/2 cup cacao powder
1/2 cup coconut nectar or maple syrup
1 teaspoon vanilla extract
Pinch of sea salt

Blend all ingredients until smooth.

COCONUT OIL CARAMEL

1 cup coconut nectar
3 tablespoons coconut oil
1 teaspoon vanilla extract
1/2 teaspoon sea salt
1/4 cup warm water, or more depending on how thick you
    prefer it

Blend all ingredients until smooth.

SWEET AND SALTY TOASTED COCONUT FLAKES

1 tablespoon coconut oil
1 cup coconut flakes
1 tablespoon coconut sugar
Pinch of sea salt, or more to taste

Heat coconut oil in a sauté pan on medium heat. Add coconut flakes, sugar, and sea salt. Sauté until flakes are light golden brown.

coconut hibiscus panna cotta

The inspiration behind the flavor of this dish comes from a skincare product. Internally and externally, the combination of coconut and hibiscus is beauty food. Hibiscus has an amazing flavor that is lemony tart and berry rich, much like a pomegranate. Hibiscus is high in vitamin C and antioxidants. It can help reduce cholesterol and manage weight loss and is also excellent for digestion. This recipe is made using the hibiscus tea found in most health food stores and markets.

2 cups young coconut meat
1/2 cup coconut water
1/4 cup coconut nectar
1 teaspoon vanilla extract
1/2 cup brewed hibiscus tea
2 tablespoons lemon juice
1/2 cup coconut oil, melted
Candied hibiscus flowers* or fresh raspberries
Pinch of sea salt

*To make candied hibiscus flowers: lightly coat fresh, non-sprayed hibiscus flowers in maple syrup and dehydrate overnight until they are crisp.

In a high-speed blender, combine all ingredients except coconut oil until completely smooth. With the blender on low speed, pour in the coconut oil until mixture is fully emulsified. Pour into desired shape or serving container. Refrigerate or freeze until firm.

Makes 4–6 servings.

O   *Serve in individual dessert cups or shot glasses or in a traditional panna cotta ramekin.*

whole food carob brownie

This recipe is a health food classic; there is something so simple and uncomplicated about the flavor that it's really hard to beat. There is a cleanliness in using only whole food ingredients—no reduced sugars or oils, just pure, whole foods. Carob used to be a much more popular health food ingredient but lost its popularity to cacao. The flavor is slightly sweet, earthy, and less bitter than cacao. Carob is also caffeine-free and has twice as much calcium as chocolate has. This recipe is easy to make, easy to alter to fit your preferences, and great for kids.

1 1/2 cups whole walnuts
1/2 cup coconut flakes
1 cup raw carob
Pinch of sea salt
2 cups date paste*
1 teaspoon vanilla
1/4 teaspoon cinnamon
1 cup raw pecans, roughly chopped

*To make date paste, blend 2 cups pitted dates with 1 cup water (or more if the dates are dry) until completely smooth.

Place walnuts in a food processor and blend on high until the nuts are finely ground. Add coconut flakes, carob, and sea salt and continue to pulse until well combined. Add date paste, vanilla, and cinnamon and process to a dough-like consistency. Stir in chopped pecans. Press into baking pan; the mixture should be approximately 2 inches thick. Place in freezer or fridge until firm. Slice before serving.

Makes approximately 1 dozen brownies.

coconut joys

In this recipe, you can form the coconut filling into perfectly round cookies using an ice cream scoop or form them into evenly sliced bars. You can also try pressing these in a cookie sheet, covering with chocolate and chopped almonds, refrigerating, and serving. You can make this dessert nut-free without the almonds if you prefer.

CHOCOLATE GLAZE

1/2 cup melted cacao butter
1/4 cup cacao powder
1/4 cup coconut nectar
Generous pinch of sea salt

Blend all ingredients until smooth.

COCONUT FILLING

2 cups shredded coconut flakes
1 tablespoon coconut butter, melted
1 tablespoon coconut oil, melted
2 tablespoons coconut nectar or white honey
Scrapings from 1 vanilla bean, or 1 teaspoon vanilla extract
A few pinches of sea salt
A few drops of stevia (optional)
1/2 cup chopped almonds (optional)

Combine 1 cup of the coconut flakes with the coconut butter in a food processor until mixture has a chunky, dough-like consistency. Add remaining ingredients and blend until mixture is thoroughly combined and holds together well.

FINAL ASSEMBLY

Press coconut filling into the bottom of a baking pan (8 inches by 8 inches is a good size). Optionally sprinkle with almonds and sea salt. Cover with chocolate glaze. Refrigerate for at least 30 minutes before serving. Slice or break into generous pieces.

Makes approximately 12–16 servings.

white honey *and* coconut yogurt ice pops

Feel free to experiment with any fruit and flavor combination when making these ice pops. The yogurt makes an excellent base for freezing because it doesn't become icy.

2 cups warm water
1/2 cup coconut yogurt
3 tablespoons white honey
Pinch of sea salt
Scrapings from 1 vanilla bean, or 1 teaspoon vanilla extract
1–2 tablespoons coconut oil (optional)
1 teaspoon cinnamon (optional)

Blend all ingredients until smooth. Pour into ice pop molds and freeze.

Makes eight 3-ounce ice pops.

skin + body

"Taking joy in living is a woman's best cosmetic."

—ROSALIND RUSSELL

We all know that beauty is only skin deep and that external attractiveness has very little to do with internal goodness or essential quality. What goes on with our skin is often a reflection of what is happening to our bodies on the inside. The skin is regularly overlooked, but it is one of the largest organs in our body. It took me a long time to understand the importance of using only natural skincare; what gets on the skin gets inside the body. I now feel repulsed when I look at the ingredients in most skincare. If ever I do use something that is less than natural, I can actually feel the toxic weight on my body; I'm now so in tune with my skin that I can tell it is dry because of stress or hormones and more hypersensitive during times when I am more emotionally sensitive. If you listen to your body, it will tell you everything that you need to know. Finding joy in life is the way to glow—so to that extent, what happens on the inside is reflected on the outside.

When it comes to DIY skincare, many of us are suckers for beautiful packaging and luxury items. At-home remedies can seem a little too grassroots. The truth is, there is so much that you can do in your home kitchen with a simple jar of coconut oil.

Despite how it may appear, the cosmetic industry is very dirty—rampant with synthetic dyes, fragrances, stabilizers, and animal cruelty. There is substantial evidence proving that what we put on our bodies can be harmful and even toxic. The irony behind much of the cosmetic industry is that it is so heavily marketed to women—yet what makes us beautiful may actually contribute to the rise of breast cancer, birth defects, and infertility.[7] Listed in the resources section are several books and websites dedicated to this subject, and I encourage you to do research. When you start to educate yourself about the toxicity that is in a simple jar of lotion, you'll probably be more empowered to consider and create what goes on your body.

Coconut oil is a cosmetic phenomenon. The options are endless, and they really work. The recipes in this chapter are very easy. Make them unique to you by familiarizing yourself with essential oils to customize the fragrance and the benefit.

WHY COCONUT OIL WORKS

Coconut oil:

- is a natural moisturizer
- tricks skin into thinking it has enough oil to slow down oil production while locking in moisture, so it's great for either dry or oily skin
- is antibacterial and antifungal, which makes it great for acne and eczema
- is non-irritating because it doesn't change the pH of your face
- is full of antioxidants to help diminish fine lines, wrinkles, and dark spots
- works as a natural SPF

coconut oil face wash

Washing your face with oil seems very counterintuitive, but it is hands down one of the greatest beauty secrets of all time. We spend our lives scrubbing our faces and removing all oils and beneficial bacteria. Many people have subscribed to the wash, tone, and moisturize regime. The beauty of coconut oil is that it completes all of these in one easy step, and it often eliminates the need for spot treatments, serums, and makeup.

1 tablespoon coconut oil
1 warm, wet washcloth; you can add an essential oil like lavender or peppermint to the washcloth to either relax or wake up your face

Massage coconut oil onto your face for at least 20 seconds. It is safe to use on your eyes to remove any eye makeup. Wipe off makeup and cleanse skin with a warm washcloth. You can rinse your face to remove excess oil, but do not wipe off all oil—leave a little on your skin and it will feel clean and glow. Use daily.

Makes 1 wash.

○ *If you have extremely oily or acne-prone skin, try adding a little tea tree, oregano, bergamot, or rosemary essential oil.*

mint chip lip balm

What goes on your lips can matter more than any area of your body. When you apply lip gloss, lipstick, or synthetic lip balm, you end up eating more than half of it. Because of the high usage of synthetic fragrances, preservatives, and petroleum-based ingredients in makeup, it is very important to consider what is in your lip products. Fortunately, safer products are regularly being seen on the market, but it is also very easy to make your own.

This recipe is very basic; feel free to substitute different oils, butters, and essential oils to your liking.

2 tablespoons cocoa butter (shea or mango butter are nice alternatives)
2 tablespoons beeswax
2 tablespoons coconut oil
2–3 drops peppermint essential oil
1 teaspoon vanilla extract

Coarsely chop or grate the beeswax and butter together. Using a double boiler or low-heat method, melt the cocoa butter and wax together; stir gently. Be very careful to not overheat or boil the product. Add coconut oil and whisk all components together. Once all ingredients are well combined, remove from heat and stir in essential oils and vanilla. Pour the mixture into lip balm containers or tins.* You can refrigerate to help solidify the ingredients more quickly.

It is very easy to buy bulk lip balm tins online. This product makes a really unique custom gift.

Makes approximately 1 dozen lip balms, depending on the size of your containers.

○ *The suggested formula is equal parts oil, butter, and wax. Beeswax from high-quality local sources is recommended, but if you prefer to make this product completely vegan, you can substitute carnauba wax, although it is a little harder to find. Wax is a necessary component that helps solidify the product so that it doesn't melt as easily.*

coconut and honey body scrub

Making your own body scrub is so easy. Sugar scrubs act differently than salt, but you can substitute salt in this recipe if you prefer. Salt makes a much harsher scrub that is better for feet, elbows, and rough patches of skin.

Sugar scrub is the gentlest type of body scrub. On top of being an excellent exfoliant, sugar has many other added benefits when applied topically. Sugar dissolves easily in water, and because of its light, sticky quality, it attaches to skin and acts as a carrier for oils and moisturizer. Sugar is a natural humectant because it draws moisture from the environment into the skin, making it a very powerful moisturizer. Sugar is a natural source of alpha hydroxy acids and helps generate cell turnover, which is excellent for improving aged or sun-damaged skin.

Coconut sugar is suggested in this recipe. Coconut sugar has a great texture, and it breaks down into very small micro beads that provide a really soft and deep exfoliation. However, coconut sugar can be a bit pricey, so feel free to use generic white sugar if you would like.

1 cup coconut sugar
1/2 cup coconut oil
2 tablespoons honey
A few drops of the essential oil of your choice*

*Rose essential oil is an excellent choice in this recipe because it compliments the moisturizing and antiaging effects of the sugar.

Mix all ingredients together. Use only a few tablespoons at a time. Make sure to keep the mixture away from water or the sugar will dissolve before it can be used. Use this three to four times a week or as often as preferred.

Makes enough for approximately 2 weeks of use.

coconut shampoo

This is a very simple shampoo recipe that uses a base of pure Castile liquid soap that can be found in all health food stores. Pure Castile soap is traditionally made from a base of olive oil and sodium hydroxide but can include any vegetable-based soap that is made without using harsh chemicals or dyes. Castile soap can be used for almost anything—dish soap, laundry detergent, house cleaner, and more. You can mix it up by changing the essential oil depending on the use. If you feel like diving in to a bit of chemistry, you can make your own soap. There are many articles and Internet resources to instruct you in this method, but buying it is much easier.

This shampoo recipe is particularly nourishing and grounding. Palo santo is a really beautiful oil to use in shampoo because of its calming qualities. Palo santo, or "holy wood," has an intoxicating, sweet incense and earthy aroma—the scent is subtle but at the same time very powerful in its healing nature. The scent, along with patchouli, mixes well with the natural musk of hair. Outside of the spiritual and meditative realms, palo santo oil has been used for cancer prevention, as an immune system stimulant, as a muscle relaxer, and to calm nerves.

1/2 cup liquid Castile soap
1 teaspoon jojoba oil
1/8 teaspoon vitamin E oil
1/4 cup coconut oil
3 tablespoons aloe vera gel
10 drops rosemary essential oil
10–15 drops palo santo essential oil (optional)
5 drops patchouli (optional)

Blend all ingredients until smooth. Allow mixture to settle a few minutes before using. You can use a few tablespoons of distilled water to help emulsify the ingredients. Store in a squeeze bottle or in a recycled shampoo bottle.

Makes enough for 10–12 uses, depending on the length and texture of your hair.

coconut conditioner

The combination of coconut oil and vitamin E is so simple and nourishing. This recipe works especially well for dry or color-treated hair. If you have naturally oily or very thin hair, you may find the oil to be a little strong; experiment with using less or more oil depending on the texture of your hair. Vitamin E and coconut oil both have beneficial antioxidants that protect and strengthen the hair shaft.

1/4 cup coconut oil, melted
1/2 teaspoon vitamin E oil
5 drops of your favorite essential oil*

*I use peppermint for a fresh, clean scent. If you are looking for something more exotic, palo santo is also really nice in this recipe.

Mix all ingredients together in a small bowl and chill.

Makes enough for 2–4 hair washes.

- If you have a dry scalp, eczema, or psoriasis, use this recipe as a hot oil scalp treatment.
- **To use as a conditioner:** After washing hair, rub 1–2 tablespoons between the palms of your hands and smooth onto the hair shaft. It is best to avoid roots; otherwise, hair can look greasy.
- **To use as a scalp treatment:** Thoroughly massage melted oil onto scalp. Leave on for least 1 hour or longer. Wash hair. Tea tree or oregano oil is excellent for scalp treatment.

coconut acne face mask *or* wash

Despite being oily, coconut oil is an excellent solution for acne or troubled skin. The anti-inflammatory and antibacterial properties make it a powerful force against breakouts. This recipe can be used as either a face wash or face mask.

2 tablespoons extra virgin coconut oil
2 teaspoons raw honey
1 teaspoon lemon juice or apple cider vinegar
A few drops of lavender essential oil
A few drops of tea tree essential oil

Mix all ingredients together. Rub into skin and immediately rinse off, or leave on skin for up to 20 minutes for a more powerful treatment.

Makes 1 treatment.

oil pulling

Oil pulling is the basic method of swishing one to two tablespoons of sesame or coconut oil in your mouth for at least twenty minutes. The prescription for this recipe is to do it on a completely empty stomach.

Although oil pulling is an ancient method of improving oral health, the benefits are just starting to be recognized in western culture. This practiced originated in Ayurvedic tradition as a means to improve dental hygiene. Oil pulling works by dissolving the toxins and bacteria in your mouth, cutting through plaque, and absorbing into your skin through your gums and tongue. Sesame oil was traditionally used, but coconut oil is preferred in this recipe because of its natural antibacterial and immune-boosting properties. There are many reported benefits behind the method of oil pulling. Some of the most easily noted improvements will include whiter teeth, stronger teeth and gums, better breath, and less gum and mouth sensitivity.

Whether or not you fully believe in the benefits, give oil pulling a try. At the very least, coconut oil makes a great organic mouthwash substitution because of its vitamin E and its antibacterial and antifungal properties.

2 cups coconut or sesame oil

Swish 1–2 tablespoons of oil around in your mouth for at least 20 minutes. Spit. Rinse mouth with warm water. Brush and floss like normal. Store the rest of the oil in a jar for future uses.

Makes one 16-ounce jar of coconut or sesame oil that provides 25–32 uses.

○   *Oil pulling is also great to do after a night of toxins because it can help alleviate hangovers.*

coconut foot balm

Making sure that your feet are well moisturized and exfoliated is important. Your feet are your grounding force, and keeping them healthy allows you to feel good and stay active. Using magnesium flakes enhances the benefits of this recipe. Magnesium flakes are a pure source of magnesium chloride and trace minerals. Magnesium chloride is absorbed through the skin and boosts magnesium levels in the body. Some of the benefits of heightened magnesium levels include the reduction of aches and pains (which is why it is excellent for overworked feet), increased energy levels, increased cardiovascular support, and reduction of tension and stress. Magnesium also aids in detoxification, which improves overall mood and wellness.

Healthy feet are a sign of good hygiene and health. This recipe helps keep your *pieds* on point.

1/4 cup shea butter, melted
1/4 cup coconut oil, melted
1 tablespoon jojoba oil
1/4 cup magnesium flakes dissolved in 2 tablespoons boiling water
A few drops peppermint essential oil
A few drops rosemary essential oil

Blend shea butter, coconut oil, and jojoba until fully emulsified. Slowly add the dissolved magnesium and blend until fully combined. Add essential oils. Pour into a small jar or tin. Cool in the refrigerator for at least 15 minutes before using. The texture should be like a solid lotion. After cleaning and exfoliating, massage balm well into feet.

Makes enough for several uses.

coconut eye cream

Our eyes and the skin around them are one of the most delicate parts of our body and deserve to be treated with care. Puffiness, redness, and fine lines are all side effects of stress and not taking care of our bodies. It's very important to keep our eyes healthy and glowing. Coffee oil is an amazing ingredient to use in skincare applications, and it is especially powerful for the eye area. Coffee oil used in eye cream helps reduce puffiness, inflammation, and redness. Externally, it does much of what it does internally—it wakes you up.

2 ounces coconut oil
1 ounce cocoa butter
1/2 teaspoon vitamin E
1/2 ounce coffee-infused oil*
A few drops of rosehip seed oil (optional)**

To make coffee-infused oil: Combine 3 tablespoons of organic coffee grinds with 1/2 cup olive oil. Warm oil on very low heat for about 20 minutes, stirring constantly. Strain out coffee grinds using cheesecloth or a fine mesh strainer. This mixture can be stored in the refrigerator for up to a month and can be used in other applications.

**Rosehip seed oil is optional in this recipe, but it makes a very beneficial addition. Rosehip seed oil is incredibly moisturizing, helps even skin tone, and fights fine lines and wrinkles.*

Melt coconut oil and cocoa butter together on a stove at very low heat. Blend all remaining ingredients to fully emulsify. Refrigerate to firm before use.

Makes enough for one month of daily use.

coconut *and* magnesium headache oil

For many of us, headaches are an inevitable part of life. Daily stress, lack of sleep, anxiety, allergies, dehydration, low blood sugar, and too much time on the computer can all cause headaches. With improvements in diet and stress management, they are possible to avoid, but when they creep up, it is best to treat them naturally. This combination of magnesium and coconut oil is a great solution to add to your temples to alleviate the pain. Magnesium oil is a transdermal mineral supplement that is easily absorbed by the skin. Many of us are magnesium deficient, so increasing magnesium intake is an added health bonus. Magnesium oil relaxes the muscles and blood vessels and helps ease tension. The combination of essential oils further increases the benefits.

1 tablespoon beeswax or carnauba wax, finely grated or chopped
4 tablespoons coconut oil
2 tablespoons jojoba oil or extra virgin olive oil
1 tablespoon magnesium oil
10–15 drops peppermint essential oil
10–15 drops lavender essential oil
10 drops chamomile essential oil (optional)

Using a double boiler or low-heat method, melt the beeswax with coconut oil and jojoba or olive oil. Stir gently and make sure not to overheat. Remove from heat and add magnesium and essential oils. Whisk ingredients together to make sure they are all well combined. Pour into the container of your choice. Refrigerate to solidify before using. To use, gently massage into temples and back of neck or any spots with tension.

Makes enough for 20+ uses.

○   *This oil also makes a great salve for dry skin.*

notes

1   Ge, Liya, Jean Wan Hong Yong, Swee Ngin Tan, Xin Hao Yang, and Eng Shi Ong. "Analysis of some cytokinins in coconut (*Cocos nucifera* L.) water by micellar electrokinetic capillary chromatography after solid-phase extraction." *Journal of Chromatography A.* 1048 (2004): 119–126.

2   Fife, Bruce. *The Coconut Oil Miracle, 5th Edition.* New York: Penguin Group, 2013.

3   Ibid.

4   Macri, Irena. "Coconut Aminos: What It Is and How to Use It." November 26, 2014. EatDrinkPaleo.com. http://eatdrinkpaleo.com.au/coconut-aminos-use/

5   Kresser, Chris. "Kefir: The Not-Quite-Paleo Superfood." 2012. ChrisKressser.com. http://chriskresser.com/kefir-the-not-quite-paleo-superfood

6   http://store.immortalityalchemy.com/products/organic-raw-royal-jelly-powder

7   The Campaign for Safe Cosmetics. "Revlon Under Fire for Cancer-Causing Chemicals in Makeup." October 25, 2013. http://safecosmetics.org/article.php?id=1162

further reading

COCONUTS

1 ChrisKresser.com. www.chriskresser.com

2 The Coconut Research Center. www.coconutresearchcenter.org

3 Dr. Mercola. www.mercola.com

4 Fife, Bruce. *Coconut Cures: Preventing and Treating Common Health Problems with Coconut*. Colorado Springs, CO: Picadilly Books, Ltd., 2005.

5 Fife, Bruce. *The Coconut Ketogenic Diet: Supercharge Your Metabolism, Revitalize Thyroid Function, and Lose Excess Weight*. Colorado Springs, CO: Picadilly Books, Ltd., 2014.

6 Fife, Bruce. *The Coconut Oil Miracle, 5th Edition*. New York: Penguin Group, 2013.

7 The Weston A. Price Foundation. www.westonaprice.org

THE COSMETICS INDUSTRY

1 The Breast Cancer Fund. www.breastcancerfund.org/reduce-your-risk/tips/choose-safe-cosmetics

2 The Campaign for Safe Cosmetics. www.safecosmetics.org

3 Hart, Jolene. *Eat Pretty: Nutrition for Beauty Inside and Out*. San Francisco, CA: Chronicle Books, LLC, 2014.

4 Malkan, Stacy. *Not Just a Pretty Face: The Ugly Side of the Beauty Industry*. Gabriola Island, BC, Canada: New Society Publishers, 2007. www.notjustaprettyface.org

5 O'Connor, Siobhan. *No More Dirty Looks: The Truth about Your Beauty Products and the Ultimate Guide to Safe and Clean Cosmetics*. Cambridge, MA: Da Capo Press, 2010.

6 Women's Voices for the Earth. www.womensvoices.org

recommended brands

EXOTIC SUPERFOODS® COCONUT WATER AND YOUNG COCONUT MEAT

This is the best resource for 100 percent raw, unpasteurized, organic coconut meat and water. This product is shipped frozen to preserve freshness. At the time of this writing, Exotic Superfoods® is the purest packaged source of coconut water and young coconut meat available.*

- www.exoticsuperfoods.com

Exotic Superfoods® products were used in creating this book.

HARMLESS HARVEST® COCONUT WATER

Harmless Harvest® is readily available, high-quality organic coconut water that can be found in most health food stores. It is processed using high-pressure pasteurization to increase shelf life. The quality and flavor are outstanding, and the company is committed to sustainable business practices.

- www.harmlessharvest.com

INVO JUICE® COCONUT WATER

Invo Juice® is another brand of high-quality, high pressure–pasteurized coconut water that can be found in many health food stores across the country. The flavor and quality of their juice are excellent, and the taste is close to fresh.

- www.invo-juice.com

NUTIVA® HEMP, COCONUT, CHIA, AND RED PALM

Nutiva® brand sources 100 percent organic and fair trade products. Their business model is focused on social and environmental sustainability and responsibility. The flavor and the quality of all of their products are superb.*

- www.nutiva.com

Nutiva® coconut butter is referred to as "coconut mana" but is a high-quality coconut butter.

EARTH CIRCLE ORGANICS® COLD-PRESSED VIRGIN COCONUT OIL, COCONUT PALM SUGAR, COCONUT FLAKES, AND COCONUT FLOUR

Earth Circle Organics® offers a wide variety of high-quality coconut oil and other coconut-related products.

- www.earthcircleorganics.com

COCONUT SECRET® COCONUT NECTAR, COCONUT CRYSTALS, COCONUT VINEGAR, COCONUT AMINOS, AND COCONUT FLOUR

At the time of this writing, Coconut Secret® brand produces the most high-quality and diverse array of coconut products made from coconut sap on the market.

- www.coconutsecret.com

COCUGAR® THAI COCONUT SUGAR

Cocugar® makes delicious, 100 percent pure organic Thai coconut sugar.

- www.cocugar.com

CULTURES FOR HEALTH® KEFIR STARTER CULTURES

Cultures for Health® is the best and highest-quality online source for kefir starter cultures.

- www.culturesforhealth.com

IMMORTALITY ALCHEMY® BEE MANA ROYAL JELLY, REISHI, AND CORDYCEPS

Immortality Alchemy® sells tonic-grade herbs and supplements sourced from the highest quality raw materials. All of their powder products are binder-, excipient-, preservative-, and additive-free. They contain no starch, dairy, wheat, corn, soy, or any material other than what is stated on the label.

- www.immortalityalchemy.com

SUN POTION® TRANSFORMATIONAL FOODS

This is an excellent source of medicinal plants, superfoods, and tonic herbs sourced from only wildcrafted and/or organic products from around the world. Located in Santa Barbara, this company is founded on holistic wellness and integrity.

- www.sunpotion.com

ROYAL SENSE® BULGARIAN ROSE WATER

Royal Sense® is the leader in authentic, 100 percent pure and natural Bulgarian rose water (rose hydrosol) for internal and external uses. Their product is handcrafted in the Rose Valley in Bulgaria and contains only pure rose extract without any sweetener or filler.

- www.royalsenseusa.com

ELIZABETH VAN BUREN® ESSENTIAL OILS

This company offers high-quality, therapeutic-grade essential oils, produced using the most advanced distillation techniques.

- www.elizabethvanburen.com

about

With a lifelong passion for a healthy lifestyle and a love of vegetarian cuisine, South Carolina native Meredith Baird has never been one to take the traditional route. During college, she studied French and art history, where she refined her passion for art, culture, and design. Following a vision to integrate good design with health and wellness, she pursued her culinary passions further to become a certified raw food chef and instructor. She believes that health and beauty are one and the same and that mind, body, spirit, and plate are all organically interconnected. Through her culinary art, she focuses on vegan cuisine, but her food philosophy has evolved over the years to represent an attitude of mindfulness and enjoyment. Her recipes are designed to inspire and let nature do most of the work.

Meredith is the co-author of *Raw Chocolate*, *Everyday Raw Detox*, and *Plant Food*. This is her fourth book.

acknowledgments

It's hard to imagine that a coconut cookbook would house so many life experiences. I suppose I shouldn't be surprised, as I have always felt intuitively that turning 30 would be a big year for me. Throughout the course of writing this book, a lot has happened to me on a personal level. I experienced loss, followed by even more incredible amounts of gain. I had to learn to let go and trust in fate. I lived alone for the first time in many years. I moved twice. I changed career paths. I made friends and I lost friends. I chopped all my hair off. I ran a half marathon . . . among other things. But most serendipitously, I found love.

Does writing a book about coconuts bring love into your life? No, of course not. But the reason the two are associated in my mind is that this book serves as a testament to the powers of attraction to focus. Do your work. Breathe deep. Smile, and no matter what, find that peace within. Don't be so focused that you forget to look up, but don't lose sight of your course. Have a purpose. Having a mission is the ultimate attraction and perhaps the key to finding happiness.

Some of the people who helped me pull this book off have been my dearest friends for a long time. Without their generosity of time, talent, and effort, this wouldn't have been possible. Simone Powers and I have served as each other's sidekicks on an endless array of projects. She is always upbeat, helps me think clearly, and makes me laugh. We are good at making food together, and she is extremely talented in her own right. Greer Inez made this book what it is by contributing her talent for photography, and she worked for very little in return. She makes me feel like the coolest person on the planet. Thank you. I appreciate endlessly the friends who believe in me and want to help me. And thanks to all of my dears who read the book and gave thorough feedback—despite most likely having other books on the reading list.

Thanks to the Familius team, who I sincerely enjoyed working with. I appreciate their diligence in editing and designing and their support in making this project come to life.

Thanks to my three-unit family, who has always supported me in my (what used to be very fringe) efforts.

And thanks to my love, who has a creative eye and artistic vision that is refined beyond measure. He worked away behind the scenes with his incredible design skills and helped make the look of this book what it is today. I learned so much.

My mission has always been to inspire people to live a better life by being thoughtful in everything I do. Food is medicine, but it is only a piece of the pie. Laughter, love, and happiness are immeasurable assets to wellness and longevity. A coconut or green juice will only get you so far. I write books because I like to write books and because I like to give people ideas. There is no gimmick involved.

Like the coconut, the older we get, the more diverse we get—the thicker the skin and the harder the shell—but, also like the coconut, we don't lose our flavor or versatility. Crack it all open by hitting it hard—it might not be easy, but it's totally worth it. The inside is the sweetest gift of all!

I look forward to entering this next chapter with grace and integrity.

conversions

Volume Measurements

U.S.	METRIC
1 teaspoon	5 ml
1 tablespoon	15 ml
1/4 cup	60 ml
1/3 cup	75 ml
1/2 cup	125 ml
2/3 cup	150 ml
3/4 cup	175 ml
1 cup	250 ml

Weight Measurements

U.S.	METRIC
1/2 ounce	15 g
1 ounce	30 g
3 ounces	90g
4 ounces	115 g
8 ounces	225 g
12 ounces	350 g
1 pound	450 g
2 1/4 pounds	1 kg

Temperature Conversion

FAHRENHEIT	CELSIUS
250	120
300	150
325	160
350	180
375	190
400	200
425	220
450	230

index

about familius

Welcome to a place where mothers and fathers are celebrated, not belittled. Where values are at the core of happy family life. Where boo boos are still kissed, cake beaters are still licked, and mistakes are still okay. Welcome to a place where books—and family—are beautiful. Familius: a book publisher dedicated to helping families be happy.

VISIT OUR WEBSITE: WWW.FAMILIUS.COM

Our website is a different kind of place. Get inspired, read articles, discover books, watch videos, connect with our family experts, download books and apps and audiobooks, and along the way, discover how values and happy family life go together.

JOIN OUR FAMILY

There are lots of ways to connect with us! Subscribe to our newsletters at www.familius.com to receive uplifting daily inspiration, essays from our Pater Familius, a free ebook every month, and the first word on special discounts and Familius news.

BECOME AN EXPERT

Familius authors and other established writers interested in helping families be happy are invited to join our family and contribute online content. If you have something important to say on the family, join our expert community by applying at:

www.familius.com/apply-to-become-a-familius-expert

GET BULK DISCOUNTS

If you feel a few friends and family might benefit from what you've read, let us know and we'll be happy to provide you with quantity discounts. Simply email us at specialorders@familius.com.

Website: www.familius.com
Facebook: www.facebook.com/paterfamilius
Twitter: @familiustalk, @paterfamilius1
Pinterest: www.pinterest.com/familius

The most important work you ever do will be within the walls of your own home.